INTERNATIONAL STANDARD UNITS FOR WATER AND WASTEWATER PROCESSES

WEF Manual of Practice No. 6

Prepared by the International Standard Units for Water and Wastewater Processes Task Force of the Water Environment Federation, the American Water Works Association, and the International Water Association

2011

Water Environment Federation
601 Wythe Street
Alexandria, VA 22314-1994 USA

American Water Works Association
6666 West Quincy Avenue
Denver, CO 80235 USA

International Water Association
Alliance House
12 Caxton Street
London SW1H 0QS
United Kingdom

International Standard Units for Water and Wastewater Processes

About WEF

Formed in 1928, the Water Environment Federation® (WEF®) is a not-for-profit technical and educational organization with members from varied disciplines who work towards WEF's vision to preserve and enhance the global water environment.

For information on membership, publications, and conferences, contact

Water Environment Federation
601 Wythe Street
Alexandria, VA 22314-1994 USA
(703) 684-2400
http://www.wef.org

About American Water Works Association

Founded in 1881, AWWA is an international nonprofit and educational society with more than 60,000 members. AWWA advances public health, safety and welfare by uniting the efforts of the full spectrum of the water community. Membership includes more than 4,600 utilities that supply drinking water to roughly 180 million people in North America.

Through our collective strength we become better stewards of water for the greatest good of the people and the environment.

For information on membership, publications, and conferences, contact

American Water Works Association
6666 West Quincy Avenue
Denver, CO 80235 USA
1-303-794-7711

Or visit our web site at http://www.awwa.org

About International Water Association

The International Water Association (IWA) is the global network of 10,000 water professionals. Through its international members and experts in research, practice, regulation, industry, consulting, and manufacturing, IWA helps water professionals create innovative, pragmatic, and sustainable solutions to challenging global needs.

For information on membership, publications, and conferences, please contact

International Water Association (IWA)
Alliance House
12 Caxton Street
London SW1H 0QS
United Kingdom
Telephone: +44 207 654 5500
Email: water@iwahq.org
http://www.iwahq.org/

Prepared by the **International Standard Units for Water and Wastewater Processes** Task Force of the **Water Environment Federation,** the **American Water Works Association,** and the **International Water Association**

Kendra Sveum, P.E., *Chair*

Jorge T. Aguinaldo
Paul Bizier
Marie S. Burbano, Ph.D., P.E.
Jennie S. Celik, P.E.
Laurie Chase, P.E.
Pat Coleman, Ph.D., P.Eng.
Michael E. Davis, Ph.D.
Bradford E. Derrick, P.E.
Mridula Deshpande, P.E.
Gil Dichter
Ernest U. Earn
Richard Edwards, P.E.
Richard Finger
Val S. Frenkel, Ph.D., P.E., D.WRE.
David Guhin, P.E.
Francis J. Hopcroft, P.E., LSP
Ray Iehl
David L. Jaeger
Hari K. Kapalavai, P.E., LEED GA
Michael W. Kirby, P.E.
Jian Li, Ph.D., P.Eng., P.E.
Maritza Marcias-Corral, Ph.D.
James J. Marx, P.E.

Jon H. Meyer
JB Neethling
James Newton, P.E., BCEE
Robert C. O'Day, CHMM, QEP
Peter Petersen
Matthew Peyton
Shane C. Porter, P.E.
Doug Prentiss
Joseph C. Reichenberger, P.E., BCEE
David Riedel
Craig Riley, P.E.
Damon K. Roth, P.E.
Nelson Schlater, P.E.
David J. Schroeder, Jr., P.E.
Kenneth Schnaars, P.E.
Jimmy Stewart
Stephen E. Sticklen, P.E.
Justyna Teper
Daniel Unger
Grant Van Hemert, P.E.
Sridhar Vedachalam
Mandeera Wagle, P.E.
Marc P. Walch, P.E.
G. Elliott Whitby, Ph.D.

Under the Direction of the **Municipal Design Subcommittee** of the **Technical Practice Committee**

2011

Water Environment Federation
601 Wythe Street
Alexandria, VA 22314–1994 USA
http://www.wef.org

American Water Works Association
6666 West Quincy Avenue
Denver, CO 80235 USA
http://www.awwa.org

International Water Association
Alliance House, 12 Caxton Street
London SW1H 0QS, UK
http://www.iwahq.org

Manuals of Practice of the Water Environment Federation®

The WEF Technical Practice Committee (formerly the Committee on Sewage and Industrial Wastes Practice of the Federation of Sewage and Industrial Wastes Associations) was created by the Federation Board of Control on October 11, 1941. The primary function of the Committee is to originate and produce, through appropriate subcommittees, special publications dealing with technical aspects of the broad interests of the Federation. These publications are intended to provide background information through a review of technical practices and detailed procedures that research and experience have shown to be functional and practical.

Contents

Preface xiii
List of Tables xv

Chapter 1 Development of a Uniform System of Units

1.0   INTRODUCTION 1
    1.1   Background 2
    1.2   Scope and Organization of Manual 2
2.0   INTERNATIONAL SYSTEM OF UNITS 3
    2.1   Base Units 3
    2.2   Derived Units 3
    2.3   Accepted Units 4
    2.4   Commonly Used Units in the Water Industry 5

Chapter 2 Accepted Representations of Units

1.0   UNITS AND SYMBOLS 7
    1.1   Units 7
    1.2   Symbols 8
2.0   PREFIXES 9
3.0   NUMERALS 11
4.0   CONVERSION, ROUNDING, AND SIGNIFICANT FIGURES 11
    4.1   Suggested References 11
    4.2   General Rules 11
    4.3   Rounding of Quantities 12

Chapter 3 Standard Units for Water Industry

1.0   WATER 13
2.0   COAGULATION 14
3.0   FLOCCULATION 15
4.0   SEDIMENTATION 15
5.0   FILTRATION 15
6.0   ADVANCED WATER TREATMENT 16
7.0   SOFTENING 17
8.0   DISINFECTION 17
9.0   EQUALIZATION STORAGE 18

10.0   DISTRIBUTION SYSTEMS 18
11.0   RESIDUALS TREATMENT 19

Chapter 4   Standard Units for Conveyance Systems

1.0   SANITARY SEWER COLLECTION SYSTEMS 21
        1.1   Flow 21
        1.2   Inflow and Infiltration 22
2.0   PLANNING/SYSTEMS ANALYSIS 22
        2.1   Pipe Design 22
        2.2   Structures 22
        2.3   Lift Stations 23
        2.4   Testing 23
3.0   STORMWATER/DRAINAGE 23
        3.1   Hydrology 23
        3.2   Pipe Systems 24
        3.3   Open Channel 25

Chapter 5   Standard Units for Wastewater Systems

1.0   WASTEWATER CHARACTERIZATION 28
        1.1   Flow 28
        1.2   Load 29
        1.3   Influent Pumping 30
        1.4   Sludge Pumping 31
2.0   PRELIMINARY TREATMENT 32
        2.1   Screening 32
        2.2   Grit Removal 32
        2.3   Odor Control 33
3.0   PRIMARY TREATMENT 33
4.0   SECONDARY TREATMENT 34
        4.1   Trickling Filters 34
        4.2   Activated Sludge 35
        4.3   Integrated Biofilm Processes 36
        4.4   Biological Aerated Filters 37
        4.5   Membrane Bioreactors 37
        4.6   Secondary Clarifiers 38
5.0   ADVANCED WASTEWATER TREATMENT 39
        5.1   Chemical Addition 39
        5.2   Chemical Precipitation 39
        5.3   Rapid Mixing 40
        5.4   Coagulation/Flocculation 40

5.5   Chemical Adsorption 41
5.6   Ion Exchange 43
5.7   Sedimentation 43
5.8   High-Rate Clarification 43
5.9   Advanced Filtration 43
5.10   Membrane Filtration 45
5.11   Gas Stripping 46
5.12   Advanced Oxidation 48
5.13   Distillation 49
5.14   Natural Treatment Systems 49
        5.14.1 Lagoons 49
        5.14.2 Land Infiltration 50
        5.14.3 Wetlands 50
        5.14.4 Water Reclamation Options 51
6.0 DISINFECTION 52
        6.1   Chlorine 52
        6.2   Sodium Hypochlorite 53
        6.3   Ultraviolet 53
        6.4   Ozone 54
7.0 SOLIDS THICKENING 55
        7.1   Gravity 55
        7.2   Dissolved Air Flotation 55
        7.3   Rotary Drums 56
        7.4   Gravity Belts 56
        7.5   Centrifuges 57
8.0 STABILIZATION 57
        8.1   Aerobic Digestion 57
        8.2   Anaerobic Digestion 58
        8.3   Lime Stabilization 59
        8.4   Composting 59
9.0 DEWATERING 59
        9.1   Belt Filter Presses 59
        9.2   Centrifugation 60
        9.3   Screw Presses 61
        9.4   Drying Beds 61
10.0 THERMAL TREATMENT 62
        10.1   Heat Drying 62
        10.2   Incineration 63
11.0 SOLIDS REUSE AND DISPOSAL 63
        11.1   Land Application 63
        11.2   Hauling and Landfilling 64

Chapter 6   Facilities

1.0   CHEMICAL SYSTEMS 66
    1.1   Basic Units 66
    1.2   Aquatic/Solution Chemistry 67
    1.3   Reaction Rates and Equilibrium Constants 68
    1.4   Chemical Handling and Feeding 68
2.0   ODOR CONTROL AND AIR EMISSIONS 69
    2.1   Odor Characterization 69
    2.2   Odor Control Technologies 70
        2.2.1   Liquid-Phase Treatment 70
        2.2.2   Biological Treatment 70
        2.2.3   Chemical–Physical Treatment Systems 71
            2.2.3.1   Gas Scrubber Systems 71
            2.2.3.2   Dry Adsorption Systems 71
        2.2.4   Combustion and Incineration 72
3.0   CONSTRUCTION QUANTITIES 72
    3.1   Data Required for Quantifying Construction Items 72
    3.2   Typical Construction Quantities 72
4.0   ELECTRICAL AND CONTROL SYSTEMS 73
    4.1   Energy Units 73
    4.2   Electrical Systems 73
        4.2.1   Direct Current Units 73
        4.2.2   Alternating Current Units 73
    4.3   Control Systems 74
        4.3.1   Memory 74
        4.3.2   Data Types 75
        4.3.3   Inputs and Outputs 75
        4.3.4   Communications 76
5.0   HEATING, VENTILATION, AND AIR CONDITIONING 76

Chapter 7   Water Reclamation and Reuse

1.0   RECLAIM, RECYCLE, OR REUSE 77
2.0   UNIQUE PARAMETERS 78

References and Suggested Readings 81
Appendix A   Conversion Table for Acceptable Units 83
Appendix B   Acronyms 89

Preface

The purpose of this Manual of Practice is to establish an international standard for units of expression that are universally understandable and readily comparable for all design, operation, and performance factors in the water and wastewater industry. Because of the importance of using a coherent system of units, the International System of Units, abbreviated to SI, has become the dominant measurement system for engineering in the international community.

The following chapters of this manual provide the SI units that are to be used in all WEF publications. The manual was written for engineering professionals familiar with water and wastewater treatment concepts, the design process, and the regulatory basis for water and wastewater control. It is not intended to be a primer for the inexperienced or the generalist.

This Manual of Practice was produced under the direction of Kendra Sveum, P.E., *Chair*. The principal authors of this Manual of Practice are as follows:

Chapter 1	Kendra Sveum, P.E.
Chapter 2	Matthew Peyton
Chapter 3	Val S. Frenkel, Ph.D., P.E., D.WRE.
	Jennie S. Celik, P.E.
	Damon K. Roth, P.E.
	Nelson Schlater, P.E.
	Daniel Unger
Chapter 4	David Guhin, P.E.
	Stephen E. Sticklen, P.E.
	Mandeera Wagle, P.E.
Chapter 5	JB Neethling
	Marie S. Burbano, Ph.D., P.E.
	Francis J. Hopcroft, P.E., LSP
	Ray Iehl
	James J. Marx, P.E.
	Justyna Teper
Chapter 6	David J. Schroeder, Jr., P.E.
	Bradford E. Derrick, P.E.
	Kendra Sveum, P.E.
	Grant Van Hemert, P.E.

Chapter 7   Craig Riley, P.E.
Appendix A   Matthew Peyton

Authors' and reviewers' efforts were supported by the following organizations:

AECOM, Newport Beach, California; Piscataway, New Jersey; Cleveland,
    Ohio; Nashville, Tennessee; and Mississauga, Ontario, Canada
Blue Heron Engineering Services, Dublin, Ohio
Calgon Carbon Corporation, Richmond Hill, Ontario, Canada
CDM, Chicago, Illinois
Chastain-Skillman, Inc., Lakeland, Florida
City of Santa Rosa Utilities Department, California
Cobb County Water System, Marietta, Georgia
Compliance EnviroSystems, LLC, Baton Rouge, Louisiana
Conestoga-Rovers & Associates (CRA), Plymouth, Michigan
Donohue & Associates, Inc., Chicago, Illinois, and Sheboygan, Wisconsin
Doosan Hydro Technology, Inc., Tampa, Florida
Doug Prentiss, Inc., Alachua, Florida
Environmental Engineering & Technology, Inc., Newport News, Virginia
HDR Engineering, Folsom, California, and Vienna, Virginia
IDEXX Laboratories, Westbrook, Maine
Kennedy/Jenks Consultants, Inc., San Francisco, California
Kent County Dept. of Public Works, Milford, Delaware
Loyola Marymount University, Los Angeles, California
Molzen-Corbin & Associates, Albuquerque, New Mexico
Orchard, Hiltz & McCliment, Inc., Livonia, Michigan
PBS&J, Orlando, Florida
San Jacinto River Authority, The Woodlands, Texas
Schneider Electric, Knightdale, North Carolina
Stantec, Windsor, Ontario, Canada
The O'Day Group, Kalamazoo, Michigan
The Ohio State University, Columbus, Ohio
U.S. Environmental Protection Agency, Chicago, Illinois
Wentworth Institute of Technology in Boston, Massachusetts
Woolpert, Charlotte, North Carolina

List of Tables

1.1	Base SI units	3
1.2	Derived SI units	4
1.3	Accepted for use with SI units	4
1.4	Commonly used units in the water industry	5
2.1	SI prefixes	9
3.1	Coagulation	14
3.2	Flocculation	15
3.3	Sedimentation	15
3.4	Filtration	15
3.5	Membrane technologies	16
3.6	Softening	17
3.7	Disinfection	17
3.8	Equalization storage	18
3.9	Distribution systems demand	18
3.10	Distribution systems design	18
3.11	Distribution systems pipe structural design	18
3.12	Distribution systems pumping	19
3.13	Residuals	19
4.1	Flow	21
4.2	Inflow and infiltration	22
4.3	Pipe design	22
4.4	Structures	22
4.5	Lift stations	23
4.6	Testing	23
4.7	Hydrology	23
4.8	Pipe systems	24
4.9	Open channel	25
5.1	Flow	29
5.2	Wastewater constituents	29
5.3	Influent pumping	31
5.4	Sludge pumping	31
5.5	Screening	32
5.6	Grit removal	32
5.7	Odor control	33
5.8	Primary clarifiers	34
5.9	Trickling filters	34

5.10   Activated sludge 35

5.11   Integrated fixed-film activated sludge and moving-bed biofilm reactors 36

5.12   Biological aerated filters 37

5.13   Membrane bioreactors 38

5.14   Secondary clarifiers 38

5.15   Chemical addition 39

5.16   Chemical precipitation 40

5.17   Rapid mixing 40

5.18   Coagulation/flocculation 40

5.19   Chemical adsorption 41

5.20   Chemical adsorption with activated carbon 42

5.21   Ion exchange 43

5.22   Advanced filtration 44

5.23   Membrane filtration 45

5.24   Gas stripping 46

5.25   Advanced oxidation 48

5.26   Distillation 49

5.27   Lagoon systems 49

5.28   Land infiltration/treatment systems 50

5.29   Wetland treatment systems 50

5.30   Water reclamation and reuse options 51

5.31   General disinfection 52

5.32   Gaseous chlorine disinfection 52

5.33   Sodium hypochlorite disinfection 53

5.34   Ultraviolet disinfection 54

5.35   Ozone disinfection 54

5.36   Gravity thickeners 55

5.37   Dissolved air flotation thickeners 55

5.38   Rotary-drum thickeners 56

5.39   Gravity-belt thickeners 56

5.40   Centrifuge thickeners 57

5.41   Aerobic digestion 57

5.42   Anaerobic digestion 58

5.43   Lime stabilization 59

5.44   Composting 59

5.45   Belt filter presses 60

5.46   Centrifugation 60

5.47   Screw presses 61

5.48   Drying beds 62

5.49   Heat drying 62

5.50   Incineration 63

5.51   Land application 64

5.52   Hauling and landfilling 64

6.1   Quantities and properties 66

6.2   Solutions 67

6.3   Reaction rates and constants 68

6.4   Chemical feeding 68

6.5   Odor characterization 69

6.6   Liquid-phase treatment 70

6.7   Biological treatment 70

6.8   Gas scrubber systems 71

6.9   Dry adsorption systems 71

6.10   Combustion and incineration 72

6.11   Typical construction quantities 72

6.12   Energy parameters and units 73

6.13   Direct current 73

6.14   Alternating current 73

6.15   Automation system memory 74

6.16   Automation system data types 75

6.17   Input and output types 75

6.18   Communications speed measurement systems 76

6.19   Heating, ventilation, and air conditioning parameters and units 76

7.1   Chemical and microbial concentrations 78

7.2   Pipeline separation 78

7.3   Irrigation 79

7.4   Groundwater recharge and recovery 79

7.5   Water rights and access 80

7.6   Reclaimed or recycled water demands and storage requirements 80

A.1   Conversion factors for acceptable units 83

Chapter 1

Development of a Uniform System of Units

1.0	INTRODUCTION		1	2.1	Base Units	3
	1.1	Background	2	2.2	Derived Units	3
	1.2	Scope and Organization of Manual	2	2.3	Accepted Units	4
2.0	INTERNATIONAL SYSTEM OF UNITS		3	2.4	Commonly Used Units in the Water Industry	5

1.0 INTRODUCTION

This manual, updated from the 3rd edition, continues its goal to establish units of expression that are universally understandable and readily comparable for all design, operation, and performance factors. All Water Environment Federation (WEF) technical papers, articles, and publications must conform to the units published in this manual. The manual was written for engineering professionals familiar with water and wastewater treatment concepts, the design process, and the regulatory basis for water and wastewater control. It is not intended to be a primer for the inexperienced or the generalist. The 4th edition of this manual is intended to reflect current design practices of water and wastewater engineering professionals. This edition includes some significant changes from the

3rd edition. Although not intended to be all-inclusive, the following list describes some of the modifications:

- Joint publication to develop an international standard,
- Inclusion of water treatment systems in addition to wastewater, and
- Removal of customary units.

1.1   Background

Historically, sanitary engineering technology was based on the measurements expressed in the foot-pound-second system (commonly known as "customary units") prevalent in the United States and most other English-speaking countries. Early editions of Manual of Practice No. 6 (MOP 6) followed this system. In 1960, however, the International Conference of Weights and Measures (Conférence Générale des Poids et Mésures [CGPM]) adopted meter-kilogram-second units as a coherent measurement system. It was logical that the Water Pollution Control Federation (now WEF), with a large and growing membership outside the United States, should be a leader in adopting and promoting metric use in the wastewater technology field. Long before similar steps were taken by other technical journals, the Federation adopted a policy in 1963 of including metric equivalents for all technical data presented in the *Journal of the Water Pollution Control Federation* (now titled *Water Environment Research*).

In the versions of MOP 6 reprinted in 1966 and 1973, metric equivalents were included. The given metric equivalent units, however, did not always conform in format or abbreviation to the style recommended in the International System of Units (abbreviated SI from the French *Le Systéme International d'Unités*). The 1973 edition of MOP 6 was drastically revised and republished in 1976 under the title "Units of Expression for Wastewater Treatment". The 1982 edition expanded the scope beyond treatment to include many other facets of wastewater management.

1.2   Scope and Organization of Manual

This manual consists of seven chapters, each of which focuses on a particular sector of the water and wastewater industry. Following is a brief overview of some of the major topics covered by each chapter:

- Chapter 1 presents the purpose and scope of the manual,
- Chapter 2 addresses rules and conventions for using and representing units,
- Chapter 3 covers units used with water treatment systems,
- Chapter 4 presents standard units for water and wastewater conveyance systems,
- Chapter 5 covers units used with wastewater treatment systems,

- Chapter 6 covers units used with facilities associated with the support of treatment systems, and

- Chapter 7 covers units used with water reuse systems.

2.0   INTERNATIONAL SYSTEM OF UNITS

The International System of Units has become the dominant measurement system for engineering in the international community. Because of the importance of using a coherent system of units, SI is heavily used throughout science- and technology-based industries. The following chapters of this manual provide the SI units that are to be used in all WEF publications. When the field of application or special needs of the user require the use of other units, the value first should be expressed in SI units; the other units may follow in parentheses.

The SI units currently are divided into base units and derived units, which together form the coherent system of SI units. In a coherent system, no numerical factor other than the number "1" ever occurs in the expressions for the derived units in terms of the base unit. The following sections also discuss some widely used non-SI units that are accepted for use with SI, although the unit system is no longer coherent.

2.1   Base Units

Chapter 2 defines the SI prefixes that are used to form multiple and submultiples of the base SI units. The use of prefixes allows for very large or very small numbers to be avoided.

TABLE 1.1   Base SI units.

Parameter	Name	Symbol
Length	meter	m
Mass	kilogram	kg
Time	second	s
Electric current	ampere	A
Thermodynamic temperature	kelvin	K
Amount of substance	mole	mol
Luminous intensity	candela	cd

2.2   Derived Units

The group of derived units is not intended to be all-inclusive, but does list those common to the water and wastewater industry. The SI prefixes may be used with any of the special names and symbols, but the resulting unit will no longer be coherent.

TABLE 1.2   Derived SI units.

Parameter	Name	Symbol
Plane angle	radian	rad
Solid angle	steradian	sr
Frequency	hertz	Hz
Force	newton	N
Pressure	pascal	Pa
Energy, work, heat	joule	J
Power	watt	W
Electric charge, quantity	coulomb	C
Electric potential difference, electromotive force	volt	V
Capacitance	farad	F
Electric resistance	ohm	Ω
Electric conductance	siemens	S
Magnetic flux	weber	Wb
Magnetic flux density	tesla	T
Inductance	henry	H
Celsius temperature	degrees Celsius	°C
Luminous flux	lumen	lm
Illuminance	lux	lx

2.3   Accepted Units

The accepted group of units is so widely used that the International Committee for Weights and Measures accepts these units for use with SI. It should be noted that the incorporation of these units makes the resulting system incoherent.

TABLE 1.3   Accepted for use with SI units.

Parameter	Name	Symbol
Time	minute	min
Time	hour	h
Time	day	d
Angle	degree	°
Angle	minute	′
Angle	second	″
Volume	liter	L
Area	hectare	ha
Mass	metric ton	t

2.4   Commonly Used Units in the Water Industry

Commonly used SI units in the water and wastewater industry are included here for easy reference.

TABLE 1.4   Commonly used units in the water industry.

Parameter	Unit	Comments
Length	mm	
Length	cm	
Length	m	
Length	km	
Area	cm^2	
Area	m^2	
Area	ha	
Area	km^2	
Mass	t	
Energy	J	
Energy	kW·h	
Power	W	
Temperature	°C	
Pressure	kPa	
Pressure	m H_2O	
Flow	ML/d	Time may need to be adjusted based on application
Flow	L/d	Time may need to be adjusted based on application
Flow	m^3/s	Time may need to be adjusted based on application
Velocity	m/s	
Loading rate	$m^3/(m^2 \cdot h)$	Volumetric loading rate
Loading rate	$kg/(m^2 \cdot d)$	Mass loading rate
Loading rate	kg/m^3	Mass loading rate

Chapter 2

Accepted Representations of Units

1.0	UNITS AND SYMBOLS	7		4.1	Suggested References	11
	1.1 Units	7		4.2	General Rules	11
	1.2 Symbols	8		4.3	Rounding of Quantities	12
2.0	PREFIXES	9				
3.0	NUMERALS	11				
4.0	CONVERSION, ROUNDING, AND SIGNIFICANT FIGURES	11				

1.0 UNITS AND SYMBOLS

1.1 Units

Unit names are shown using roman (regular) type font, regardless of the type used in the surrounding text (International Bureau of Weights and Measures, 2006). Unit names, including prefixes, are treated as nouns and are not capitalized when spelled out. The exceptions to this rule are units at the beginning of a sentence, in titles and headings, and in other instances in which all other main words are capitalized. Another exception is the unit "degrees Celsius" where "Celsius" is always capitalized.

A common misconception is that if a unit is derived from a person's name, that the unit should be capitalized. Only units that correspond to the above exceptions should be capitalized.

7

> **Example:** the units "hertz" and "pascal" are derived from a person's name and should not be expressed as "Hertz" or "Pascal".

Terms that indicate a unit is raised to a power such as "squared" or "cubed" should be placed after a unit name. However, when expressing a unit of area or volume raised to a power, the terms "square" and "cubic" should be placed before the unit name (International Bureau of Weights and Measures, 2006).

> **Example:** square meter (m^2) is an area, cubic millimeter (mm^3) is a volume, kilometer per second squared (km/s^2) is neither an area nor a volume.

Spelling should be "liter" and "meter" for WEF publications, although the spellings "litre" and "metre" will remain widespread outside of the United States.

1.2   Symbols

Unit symbols are shown using roman (upright) type font, regardless of the type used in the surrounding text (International Bureau of Weights and Measures, 2006). The symbols of SI units are not capitalized unless the unit was derived from a person's name, in which case the first letter of the symbol is capitalized. This rule applies only to "unit symbols"; refer to Section 1.1 for capitalization of unit names. The unit "liter" is an exception, the symbol should be "L" to avoid possible confusion with the number one "1" (International Bureau of Weights and Measures, 2006).

> **Example:** the unit symbols for newton (N) and joule (J) should not be expressed as "n" or "j".

Unit symbols are not followed by a period unless they are located at the end of a sentence. Unit symbols should never be used as abbreviations in text; if the unit symbol is not preceded by a numerical value, the entire unit name should be used.

> **Example:** The following statements correctly illustrate how units and symbols should be represented in text: "The holding tank's capacity is expressed in cubic meters" (unit name is used); or "The capacity of the holding tank is 3000 m^3" (numerical value precedes unit symbol). The statement "The holding tank's capacity is expressed in m^3" uses a standalone symbol as an abbreviation, which is incorrect.

When multiplying units, the product should be expressed with either a space or a half centered dot (·) separating the symbols to avoid confusion with a unit prefix.

Example: N m or N·m for a newton meter.

The quotients of unit symbols should be expressed using an oblique stroke (/), a horizontal line, or a negative exponent. Brackets should be used to avoid confusion when combining several units in the quotient. If using a negative exponent, then the rules of symbol multiplication should be followed (International Bureau of Weights and Measures, 2006).

Example: m/s, $\frac{m}{s}$, m s^{-1}, or m·s^{-1} for meters per second; m kg/(s^3 A) or m kgs^{-3} A^{-1} for volt per meter.

Units of time should always be last when expressed in the numerator or denominator of a quotient.

Example: m^3/(ha·d), not m^3/(d·ha) when expressing cubic meter per hectare per day.

2.0 PREFIXES

The base and derived units shown in Chapter 1 of this manual can be used with the addition of prefixes to indicate larger and smaller measures.

Prefixes are shown using roman (regular) type font regardless of the type used in the surrounding text (International Bureau of Weights and Measures, 2006). Capitalization of prefix symbols is important when using SI units, and

TABLE 2.1 SI prefixes.

Multiplier	Prefix	SI Symbol
1 000 000 000 = 10^9	giga	G
1 000 000 = 10^6	mega	M
1000 = 10^3	kilo	k
100 = 10^2	hecto[a]	h
10 = 10^1	deca[a]	da
0.1 = 10^{-1}	deci[a]	d
0.01 = 10^{-2}	centi[b]	c
0.001 = 10^{-3}	milli	m
0.000 001 = 10^{-6}	micro	μ
0.000 000 001 = 10^{-9}	nano	n

[a]Use should be avoided.

[b]For nontechnical use only, except "centimeter".

must be used exactly as shown in the table above to avoid confusion. Use of the following SI prefixes should be avoided: hecto (hectare is an exception), deka, deci, and centi (centimeter is an exception). The most commonly used prefixes are in powers of three. The complete prefix should be used; exceptions include megohm, kilohm, and hectare—three cases where the final vowel is omitted. Typically, both vowels are retained and are pronounced.

> **Example:** "milliampere" and "gigaohm".

In all cases of prefix use (including symbols), a space or hyphen is never used to separate the prefix from the unit.

> **Example:** millimeter and kPa; not milli-meter, milli meter, k-Pa, or k Pa.

Select the unit prefix that makes the most sense for the value of the number. Typically, if the value of a number exceeds 1000 or is reported to the one thousandth decimal place, the next larger or smaller prefix is used, respectively.

> **Example:** 5 mm or 50 mm instead of 0.005 m or 0.05 m; 100 km instead of 100 000 m.

A prefix used in a table or discussion should be consistent throughout, even if this requires five or six digits before the decimal point. If the chosen prefix is millimeters then this prefix should be used whether it is 0.5 mm or 50 000 mm. The mixture of prefixes should be avoided unless the difference in size is extreme.

> **Example:** 1500 m of 2-mm wire is acceptable; 1 500 000 mm of 2-mm wire is not acceptable.

Double prefixes should never be used; the correct use is "milligram", not "microkilogram". Slang terms should not be used in technical writing. For example, do not substitute "kilo" for "kilogram".

When listing a gas volume that is to be measured at a standard temperature and pressure (20 °C and 101.325 kPa and 36% relative humidity), use N for normalized before the unit. To distinguish between N for newton and N for normalized there is no multiplication "dot" between the N for normalized and the unit but will be treated as a prefix. If the normalized unit in the manual is measured at other conditions than what is listed above, they will be included in the comments column for reference.

> **Example:** normal cubic meter (Nm^3) and newton meter ($N \cdot m$)

3.0   NUMERALS

Numbers with five or more digits should be separated into groups of three by a small space; this is equally applicable to whole numbers and decimals. When a number has four digits on either side of the decimal marker, it is customary not to use a space (International Bureau of Weights and Measures, 2006). The use of commas as a separator is not acceptable for use with SI units. Only a period should be used to differentiate between whole numbers and decimals. When expressing numbers less than one, a zero should always precede the decimal point.

> **Example:** 5 845 254 286 is acceptable; 5,845,254,286 is not acceptable; 0.2879 and 0.258 624 125 are acceptable; .2879 and 0.258,624,125 are not acceptable; 534 879.231 456 is acceptable; 534.879,231.456 and 534,879.231,456 are not acceptable; 2975.1843 is customary; 2 925.184 3 is not customary.

4.0   CONVERSION, ROUNDING, AND SIGNIFICANT FIGURES

4.1   Suggested References

The *Guide for the Use of the International System of Units (SI)* is a thorough reference tool for converting standard units to SI. Appendix A contains many conversion factors and can be downloaded from the National Institute of Standards and Technology (NIST) at www.nist.gov. In addition, common conversion factors are included in Table A.1 of Appendix A of this manual.

4.2   General Rules

To convert standard units to SI, the unit value must be multiplied by a corresponding conversion factor. Although many conversion factors express multiple significant digits, the accuracy of the converted unit cannot be greater than the original. Therefore, more significant digits cannot be used in the result than included in the original number.

> **Example:** to convert from mgd to ML/d, multiply by 3.785 as in the following

- 15 mgd = 56.775 ML/d, rounded to 57 ML/d because there are two significant digits in the original number.
- 15.0 mgd = 56.775 ML/d, rounded to 56.8 ML/d because there are three significant digits in the original number.

If the number following the last significant digit is less than five, then the last significant digit is unchanged. If the number following the last significant digit is five or greater, the last significant digit is increased by 1.

> **Example:** 5.854 251 rounded to three significant digits is 5.85; 4.159 rounded to two significant digits is 4.2.

4.3   Rounding of Quantities

Converting numerical quantities requires that the maximum rounding error be taken into account in addition to the rules presented in Section 4.2. The following is an example of a situation in which the result of a conversion can have a greater number of significant digits than the original number:

> **Example:** to convert 514 gal to m^3, multiply by 3.785 412 E-03; either 514 gal = 1.946 m^3 or 514 gal = 1.95 m^3 are acceptable.

The first number, 1.946 m^3, has more significant digits than the original number. However, 514 has a rounding error of 0.0973% (0.5/514 = 0.0973%). Following the rule of Section 4.3 of this chapter, the rounding error of the result would equal 0.256% (0.005/1.95 = 0.256%), which is an increase of the rounding error. Using three significant digits would cause a loss of accuracy contained in the original number. Increasing the number of significant digits in the result to four would equate to a rounding error of 0.025 69%, decreasing the rounding error and representing more accuracy than the original number contains. In this case, the number of significant digits can be either three or four. Experience and best judgment should be used to determine the most appropriate level of accuracy to include in the converted value Thompson and Taylor, 2008).

Chapter 3

Standard Units for Water Industry

1.0	WATER	13	7.0	SOFTENING	17
2.0	COAGULATION	14	8.0	DISINFECTION	17
3.0	FLOCCULATION	15	9.0	EQUALIZATION STORAGE	18
4.0	SEDIMENTATION	15			
5.0	FILTRATION	15	10.0	DISTRIBUTION SYSTEMS	18
6.0	ADVANCED WATER TREATMENT	16	11.0	RESIDUALS TREATMENT	19

1.0   WATER

This chapter covers standard SI used for water, water treatment, and water management. The tables included here represent the significant parameters associated with water treatment and include the technological processes with the widely adopted SI units. It is broken down by the following technologies and applications:

- Coagulation—Table 3.1.
- Flocculation—Table 3.2.
- Sedimentation—Table 3.3.

- Filtration units (gravity filters, pressure filters)—Table 3.4.

- Advanced water treatment/membrane technologies (microfiltration, ultrafiltration, nanofiltration, reverse osmosis, electrodialysis, electrodialysis reversal)—Table 3.5.

- Ion exchange (ion exchange, resin softening, chemical precipitation/cold lime softening)—Table 3.6.

- Disinfection (liquid chlorine, gaseous chlorine, chlorine dioxide, UV, ozone)—Table 3.7.

- Equalization storage (volume of storage, water demand, contact time, chlorine residual)—Table 3.8.

- Distribution systems/demand [water demand, per capita demand, water loss (flow per length of pipe per pipe diameter), fire demand]—Table 3.9.

- Distribution systems/design (pipe diameter, headloss, hydraulic grade line, energy grade line, residence time, chlorine residual)—Table 3.10.

- Distribution systems/pipe structural design (load on pipe, water hammer, thrust force, soil bearing capacity, soil frictional resistance)—Table 3.11.

- Distribution systems/pumping stations (capacity, total dynamic head)—Table 3.12.

- Residuals—Table 3.13.

2.0   COAGULATION

TABLE 3.1   Coagulation.

Parameter	SI	Comments
Contact time	min	
Contaminant concentration	mg/L or µg/L	
Dose of coagulant	mg/L	
Particle counts	#/mL	
Specific UV light absorbance	L/(mg·cm)	
Turbidity	NTU	NTU = nephelometric turbidity units
Velocity gradient	s^{-1}	
Zeta potential	mV	

3.0   FLOCCULATION

TABLE 3.2   Flocculation.

Parameter	SI	Comments
Flocculation time/contact time	min	
Velocity gradient	s^{-1}	Derived from velocity (m/s) divided by distance (m)
Dose of flocculant	mg/L	
Hydraulic retention time	min	

4.0   SEDIMENTATION

TABLE 3.3   Sedimentation.

Parameter	SI	Comments
Particle settling rate	m/h	
Solids collector velocity	m/min	
Solids loading	$kg/(m^2 \cdot d)$	
Surface overflow rate	m/d	Can be expressed as $m^3/m^2 \cdot d$
Volume settled solids	m^3/ML	
Weir loading rate	$ML/(m \cdot d)$	

5.0   FILTRATION

TABLE 3.4   Filtration.

Parameter	SI	Comments
Air scouring airflow	$m^3/(m^2 \cdot min)$	
Backwash velocity	m/s	
Backwash time	min	
Depth of filter media	m	
Effective size	mm	
Free board of media filter	m	
Headloss	$m\ H_2O$	
Hydraulic loading rate	$L/(m^2 \cdot s)$	
Rate of rise during backwash	cm/s	
Ripening time	min	

(continued)

TABLE 3.4   Filtration (*continued*).

Parameter	SI	Comments
Run time	h	
Surface wash water pressure	kPa	
Treatment capacity	m^3/d	
Uniformity coefficient	mm/mm	
Unit filter run volume	L/m^2	

6.0   ADVANCED WATER TREATMENT

TABLE 3.5   Membrane technologies.

Parameter	SI	Comments
Chemical dose	mg/L	
Diffusivity	m^2/s	
Element dimensions	cm	
Element weight	kg	
Flowrate	m^3/d	
Flux	$L/(m^2 \cdot h)$ (LMH)	
Membrane area	m^2	
Membrane system recovery	%	
Molecular weight cutoff	daltons	
Packing density	m^2/m^3	Membrane surface area per membrane module volume
Permeability	$L/(m^2 \cdot h)/kPa$ LMH/kPa	
Pressure, pressure drop, osmotic pressure	kPa	
Salt passage	%	
Salt rejection	%	
Specific energy	kWh/m^3	
Surface area	m^2	
Transmembrane pressure	kPa	

7.0   SOFTENING

TABLE 3.6   Softening.

Parameter	SI	Comments
Alkalinity	mg/L as $CaCO_3$	
Backwash rate	m/h	
Chemical dosage	mg/L, kg/d	
Depth of resin	mm	
Detention time	min	
Dry contact surface area	m^2/g	
Empty bed contact time	min	
Exchange capacity	kg/m^3, $grains/m^3$	
Flowrate	m^3/d	
Hardness	mg/L as $CaCO_3$	
Hydraulic loading rate	$L/(m^2 \cdot s)$	
Saturation capacity	meq/mL	
Volume	m^3	
Working capacity	kg/m^3	

8.0   DISINFECTION

TABLE 3.7   Disinfection.

Parameter	SI	Comments
Chemical disinfectant dosage	mg/L	
Contact time	min or sec	
CT – Concentration × time	$(mg/L) \cdot min$	
Temperature	°C	
Chemical disinfectant residual	mg/L	
Chemical pump rate	L/min	
Chemical storage tank	m^3	
UV dose	mJ/cm^2 or $mW\text{-}s/cm^2$	
UV intensity	mW/cm^2	
UV exposure time	s	
UV power	W	
UV transmittance	%	
UV energy	J	

9.0   EQUALIZATION STORAGE

TABLE 3.8   Equalization storage.

Parameter	SI	Comments
Volume of storage	ML	
Water demand	L/d	

10.0   DISTRIBUTION SYSTEMS

TABLE 3.9   Distribution systems demand.

Parameter	SI	Comments
Water demand	L/d	
Per capita demand	L/cap·d	
Water loss	L/(m·mm·d)	
Fire demand	L/d	

TABLE 3.10   Distribution systems design.

Parameter	SI	Comments
Pipe diameter	mm	
Head loss	mm or m	
Hydraulic grade line	m	
Energy grade line	m	
Residence time	d	
Chlorine residual	mg/L	

TABLE 3.11   Distribution systems pipe structural design.

Parameter	SI	Comments
Load on pipe	kPa or kN/m^2	
Water hammer	Pa	
Thrust force	kN	
Soil bearing capacity	kN/m^2	
Soil frictional resistance	kN	

TABLE 3.12   Distribution systems pumping.

Parameter	SI	Comments
Capacity	L/d or ML/d or m^3/d	
Total dynamic head	m	

11.0   RESIDUALS TREATMENT

TABLE 3.13   Residuals.

Parameter	SI	Comments
Belt press loading	kg/(m·h)	
Cake solids concentration	mass %	Example: use in text as "cake solids concentration is 20%" to indicate 0.2 kg solids/kg cake
Capillary suction time	s	
Capture	mass %	Example: use in text as "capture was 98%" to indicate 0.98 kg of solids were retained per kg treated
Centrifuge solids throughput	kg/h	
Drying bed yield	kg/(m^2·yr)	
Freezing time	h	
Hydraulic loading rate	L/(m^2·s)	
Net pan evaporation	cm/month	
Particle settling rate	m/h	
Polymer feed rate (dry basis)	mg/kg	
Recycle rate	volume %	Example: use in text as "recycle rate was 5%" to indicate an instantaneous recycle flow rate of 0.05 m^3/d/m^3/d total plant influent flow
Sludge layer thickness	m	
Sludge production rate	kg/d	
Sludge solids concentration	mass %	Example: use in text as "sludge solids concentration was 2%" to indicate 0.02 kg solids/kg sludge
Solids collector velocity	m/min	
Solids loading rate	kg/(m^2·d)	
Specific resistance	m/kg	

(continued)

TABLE 3.13   Residuals (*continued*).

Parameter	SI	Comments
Surface overflow rate (upflow velocity)	m/d	
Time to filter	min	
Total dissolved solids	mg/L	
Total suspended solids	mg/L	
Volume settled solids	m^3/ML	
Weir loading rate	ML/(m·d)	

Chapter 4

Standard Units for Conveyance Systems

1.0	SANITARY SEWER COLLECTION SYSTEMS	21
	1.1 Flow	21
	1.2 Inflow and Infiltration	22
2.0	PLANNING/SYSTEMS ANALYSIS	22
	2.1 Pipe Design	22

2.2	Structures	22
2.3	Lift Stations	23
2.4	Testing	23
3.0	STORMWATER/DRAINAGE	23
	3.1 Hydrology	23
	3.2 Pipe Systems	24
	3.3 Open Channel	25

1.0 SANITARY SEWER COLLECTION SYSTEMS

1.1 Flow

TABLE 4.1 Flow.

Parameter	SI	Comment
Average daily flow	m^3/s	
Peak hourly flow	m^3/s	
Peak instantaneous flow	m^3/s	
Maximum daily flow	m^3/s	
Design flow	m^3/s	
Per capita flowrate	$m^3/(cap \cdot d)$	Typical: lpcd

1.2   Inflow and Infiltration

TABLE 4.2   Inflow and infiltration.

Parameter	SI	Comment
Allowable inflow and infiltration rate	$m^3/d/m^2$	
Runoff	mm/h	Typical: mm/h
Amount rain	m	Typical: mm
Amount snow	m	Typical: mm
Amount total	m	Typical: mm
Intensity	mm/h	Typical: mm/h
Duration	min	

2.0   PLANNING/SYSTEMS ANALYSIS

2.1   Pipe Design

TABLE 4.3   Pipe design.

Parameter	SI	Comment
Diameter	m	Typical: mm
Area	m^2	
Velocity	m/s	
Slope	m/m	
Hydraulic radius	m	
Elevation	m	
Depth	m	
Pressure rating	$kg \cdot m/(m^2 \cdot s^2)$	Typical: Pa or N/m^2
Head loss	m	
Roughness factor	unitless	

2.2   Structures

TABLE 4.4   Structures.

Parameter	SI	Comment
Diameter	m	
Elevation	m	

2.3   Lift Stations

TABLE 4.5   Lift stations.

Parameter	SI	Comment
Capacity	m^3/d	Typical: L/s
Pressure	$kg \cdot m/(m^2 \cdot s^2)$	Typical: Pa or N/m^2
Total dynamic head	m	

2.4   Testing

TABLE 4.6   Testing.

Parameter	SI	Comment
Hydrostatic pressure	$kg \cdot m/(m^2 \cdot s^2)$	Typical: Pa or N/m^2
Vacuum	$kg \cdot m/(m^2 \cdot s^2)$	Typical: Pa or N/m^2
Smoke	m^2/defect	Area tributary to defect
	m^3/defect	Flow tributary to defect; typical: L
	m	Length inspected
Closed circuit	m^2/defect	Area tributary to defect
Closed-circuit television	m^3/defect	Flow tributary to defect; typical: L
	m	Length inspected

3.0   STORMWATER/DRAINAGE
3.1   Hydrology

TABLE 4.7   Hydrology.

Parameter	SI	Comment
Area	m^2	
Drainage/catchment area	ha	"A" in rational equation
Frequency of occurrence	1/yr	
Rainfall intensity	mm/h	"i" in rational equation
Rainfall depth	mm	
Runoff coefficient (C)	unitless	Rational equation: $Q = CiA$
Initial abstraction	mm	
Permeability	m^2	
Soil hydraulic conductivity	mm/s	

(continued)

TABLE 4.7   Hydrology (*continued*).

Parameter	SI	Comment
Transmissivity	m^2/d	
Porosity	unitless	$\varphi = V_v/V_T$
Soil saturation	unitless	Portion of void space occupied by water
Suction head	mm	
Infiltration rate	mm/h	
Stream length	km	
Length of overland flow	m	
Slope	m/m	
Duration	h	
Time of concentration	min	
Volume	m^3	
Flowrate	m^3/s	
Velocity	m/s	

3.2   Pipe Systems

TABLE 4.8   Pipe systems.

Parameter	SI	Comment
Flowrate	m^3/s	
Velocity	m/s	
Area	m^2	
Diameter	mm	
Length	m	
Hydraulic radius	m	
Wetted perimeter	m	
Slope	m/m	
Depth	mm	
Time of concentration	min	
Travel time	min	
Mannings roughness coefficient	unitless	
Darcy friction factor	unitless	Also known as Darcy-Weisbach friction factor
Water surface elevation	m	
Specific energy	m	
Head/headloss	m	
Loss coefficient	unitless	

(continued)

TABLE **4.8**   Pipe systems (*continued*).

Parameter	SI	Comment
Acceleration from gravity (g)	m/s^2	$g = 9.81$ m/s^2 on Earth
Gutter spread	m	
Grade and cross slope	percent or %	
Inlet efficiency	unitless	
Inlet length/width	mm	
Inlet opening area	m^2	
Frequency	yr	

3.3   Open Channel

TABLE **4.9**   Open channel.

Parameter	SI	Comment
Flowrate	m^3/s	
Velocity	m/s	
Area	m^2	
Width (water surface)	m	
Depth	m	
Channel length	m	
Hydraulic radius	m	
Wetted perimeter	m	
Slope	m/m	
Time of concentration (small)	min	Applicable to ditches and small channels
Time of concentration (large)	h	Applicable to rivers and streams
Travel time	min	
Mannings roughness coefficient	unitless	
Darcy friction factor	unitless	Also known as Darcy-Weisbach friction factor
Water surface elevation	m	
Specific energy	m	
Head/headloss	m	
Loss coefficient	unitless	
Acceleration from gravity (g)	m/s^2	$g = 9.81$ m/s^2 on Earth

Chapter 5

Standard Units for Wastewater Systems

1.0	WASTEWATER CHARACTERIZATION	28		4.5	Membrane Bioreactors	37
	1.1 Flow	28		4.6	Secondary Clarifiers	38
	1.2 Load	29	5.0	ADVANCED WASTEWATER TREATMENT	39
	1.3 Influent Pumping	30		5.1	Chemical Addition	39
	1.4 Sludge Pumping	31		5.2	Chemical Precipitation	39
2.0	PRELIMINARY TREATMENT	32		5.3	Rapid Mixing	40
	2.1 Screening	32		5.4	Coagulation/ Flocculation	40
	2.2 Grit Removal	32		5.5	Chemical Adsorption	41
	2.3 Odor Control	33		5.6	Ion Exchange	43
3.0	PRIMARY TREATMENT	33		5.7	Sedimentation	43
4.0	SECONDARY TREATMENT	34		5.8	High-Rate Clarification	43
	4.1 Trickling Filters	34		5.9	Advanced Filtration	43
	4.2 Activated Sludge	35		5.10	Membrane Filtration	45
	4.3 Integrated Biofilm Processes	36				
	4.4 Biological Aerated Filters	37				

	5.11	Gas Stripping	46		7.5 Centrifuges	57
	5.12	Advanced Oxidation	48	8.0	STABILIZATION	57
	5.13	Distillation	49		8.1 Aerobic Digestion	57
	5.14	Natural Treatment Systems	49		8.2 Anaerobic Digestion	58
		5.14.1 Lagoons	49		8.3 Lime Stabilization	59
		5.14.2 Land Infiltration	50		8.4 Composting	59
		5.14.3 Wetlands	50	9.0	DEWATERING	59
		5.14.4 Water Reclamation Options	51		9.1 Belt Filter Presses	59
6.0	DISINFECTION		52		9.2 Centrifugation	60
	6.1	Chlorine	52		9.3 Screw Presses	61
	6.2	Sodium Hypochlorite	53		9.4 Drying Beds	61
	6.3	Ultraviolet	53	10.0	THERMAL TREATMENT	62
	6.4	Ozone	54		10.1 Heat Drying	62
7.0	SOLIDS THICKENING		55		10.2 Incineration	63
	7.1	Gravity	55	11.0	SOLIDS REUSE AND DISPOSAL	63
	7.2	Dissolved Air Flotation	55		11.1 Land Application	63
	7.3	Rotary Drums	56		11.2 Hauling and Landfilling	64
	7.4	Gravity Belts	56			

1.0   WASTEWATER CHARACTERIZATION

Wastewater characterization is used to describe wastewater at any point in a system (i.e., influent, throughout processes, effluent, etc.).

1.1   Flow

Flow is typically measured for the whole plant or in a specific pipeline.

TABLE 5.1 Flow.

Parameter	SI	Comment
Pipeline	m³/s	
Plant	ML/d	Typically for systems larger than 0.05 ML/d
Plant	L/d	Typically for systems smaller than 50 000 L/d
Pump	L/min	
Metering devices		
Flume throat	m	
Orifice opening	mm	
Weir length	m	

1.2 Load

Wastewater constituents typically are described as both concentrations and mass loadings. For most of the items listed (i.e., chemical oxygen demand, solids), the unit is applied for all fractions. This includes, but is not limited to, the measurement of fractions such as volatile, dissolved, colloidal, suspended, readily biodegradable, organic, and dissolved organic. Units of chemical oxygen demand (COD; i.e., mg COD/L) apply to organisms, including heterotrophs, autotrophs, phosphorus-accumulating organisms (PAOs), and others. Chemical oxygen demand units also apply to volatile fatty acids (VFAs).

TABLE 5.2 Wastewater constituents.

Parameter	SI	Comment
Alkalinity	mg/L as $CaCO_3$	Multiply by 50 to convert meq/L to mg/L as $CaCO_3$
Alkalinity	meq/L	
Aluminum	mg/L	
Bacterial counts	MPN/100 mL	Most probable number per liter
Biochemical oxygen demand		Includes all biochemical oxygen demand (BOD) fractions; 5 days, 20 °C, unless otherwise stated
Concentration	mg/L	
Load	kg/d	
Chemical oxygen demand		Includes all COD fractions
Concentration	mg O_2/L	
Load	kg/d	
Chlorine residual	mg/L as Cl_2	
Dissolved oxygen	mg/L as O_2	
Hardness	mg/L as $CaCO_3$	
Iron	mg/L as Fe	

(continued)

TABLE 5.2   Wastewater constituents (*continued*).

Parameter	SI	Comment
Micropollutants	ng/L	Includes N-nitrosodimethyl-amine, pharmaceuticals, etc.
Nitrogen compounds		Includes all nitrogen fractions
Concentration	mg/L as N	
Load	kg/d as N	
Organic carbon		Includes total and dissolved
Concentration	mg/L as C	carbon
Load	kg/d as C	
pH (standard units)	—	
Phosphorus compounds		Express all phosphorus fractions
Concentration	mg/L as P	as P
Load	kg/d as P	
Silica	mg/L as SiO_2	
Suspended solids (specify total, volatile)		Includes total suspended, volatile, organic, inorganic
Concentration	mg/L	Specify compound
Load	kg/d	
Temperature	°C	
Total Kjeldahl nitrogen		Organic nitrogen plus ammonia
Concentration	mg/L as N	nitrogen
Load	kg/d as N	
Total nitrogen		Total of organic-N, ammonia-N,
Concentration	mg/L as N	nitrate-N, nitrite-N; excludes
Load	kg/d as N	dissolved N_2
Total solids		
Concentration	mg/L	
Load	kg/d	

When expressing chemicals, the chemical species used as the basis for expression shall be included. For example,

- Ferric chloride addition shall be clarified as [mg Fe/L] or [mg $FeCl_3$/L] and
- Nitrate shall be clarified as [mg N/L] or [mg NO_3/L].

1.3   Influent Pumping

Influent pumping includes large pumping stations used to move liquid. Large and small treatment facilities require pumping at influent or intermediate stations. Several different styles of pumps are used including centrifugal, positive displacement, or screw pumps. Various arrangements of pumping stations can be used with dry or wet well arrangements. The flow entering a treatment plant typically is related to the population served and expressed in volumetric flow per day. Flow entering a treatment plant is, therefore, best described in terms of daily flow (megaliters per day). Pumps typically are sized based on instantaneous flow and, therefore, are rated on a shorter time frame, such as a liter per minute (L/min).

TABLE 5.3   Influent pumping.

Parameter	SI	Comment
Density, ρ	kg/m^3	
Dynamic viscosity, μ	mPa·s	
Efficiency	%	
Hydraulic losses	M	Friction, velocity, secondary
Kinematic viscosity, ν	μm^2/s	m^2/s $\times$ 10^{-6}
Net pump suction head	M	
Pump capacity	ML/d	
Pump capacity	L/min	
Pump head	M	Total dynamic head when pump is running
Pump power	kW	
Pump speed	rev/min	
Specific weight, γ	kN/m^3	

1.4   Sludge Pumping

Sludge pumping includes all conditions where liquids with high concentrations of suspended solids are pumped. Solids handling pumps often operate intermittently and at much lower flowrates than liquid pumps. Sludge pumps, such as return activated sludge (RAS) or backwash, pump at a higher rate similar to liquid pump criteria.

TABLE 5.4   Sludge pumping.

Parameter	SI	Comment
Density, ρ	kg/m^3	
Dynamic viscosity, μ	mPa·s	
Efficiency	%	
Kinematic viscosity, ν	μm^2/s	m^2/s $\times$ 10^{-6}
Pump capacity	ML/d	
Pump capacity	L/min	
Pump head	M	Total dynamic head when pump is running
Pump power	kW	
Solids concentration	%	mass/mass basis
Specific weight, γ	kN/m^3	

2.0   PRELIMINARY TREATMENT

2.1   Screening

Screens are used to remove large objects from wastewater to protect pumps and downstream equipment from failing and to reduce maintenance on the equipment. The screen opening defines the particle exclusion size and is chosen to suit the specific application point. Coarse screens (greater than 25-mm openings) are used to remove large objects and sometimes referred to as bar racks. Fine screens remove small particles (less than 2 mm) and typically are used upstream of membrane bioreactors.

TABLE 5.5   Screening.

Parameter	SI	Comment
Approach velocity	m/s	Based on velocity without screen
Flow velocity	m/s	Based on velocity with screen

Bar size
    Depth
    Width
	
	mm
	mm	

Clear spacing between bars	mm	
Opening	mm	Typically slots
Perforation size	mm	Typically holes
Slope from vertical	deg	
Volume of screenings	m^3/ML	Alternative: m^3/m^3

2.2   Grit Removal

Grit is removed from wastewater to protect and reduce maintenance on equipment and to prevent accumulation of grit in channels, wet wells, and other process units. Most grit removal systems rely on gravity separation of particles. Sometimes removal is enhanced by adding mechanical mixing (swirling or aeration). Key design criteria include surface loading rate and hydraulic retention time.

TABLE 5.6   Grit removal.

Parameter	SI	Comment
Approach velocity	m/s	Velocity entering unit
Channel aeration	$Nm^3/(m \cdot s)$	Equipment specific; airflow expressed in normal m^3
Detention time	min	
Horizontal velocity	m/s	Velocity through unit
Length, width, diameter	m	

(continued)

TABLE 5.6   Grit removal (*continued*).

Parameter	SI	Comment
Particle density (in water)	kg/m^3	
Particle diameter	mm	
Settling velocity for removal	m/min	Alternative: mm/s; depending on particle size
Surface loading rate	ML/m^2·d	
Vortex rotation	rev/min	
Water depth	m	

2.3   Odor Control

Odor control is provided to various process units as needed to eliminate fugitive odorous gases from escaping into the atmosphere. Odorous gases include hydrogen sulfide, amine, organic, ammonia, or other compounds. Odor control approaches include chemical feed, gas adsorption, biological beds, granular activated carbon, and others.

TABLE 5.7   Odor control.

Parameter	SI	Comment
Air flow rate	Nm^3/h	Airflow expressed in normal m^3
Detention time	min	
Dilution to threshold	mol/m^3·d	
Odor concentration	μg/m^3	Express in terms of vented gas
Odor intensity	μ mol/mol	Part per million (mol:mol)
Odor intensity	vol ppm	Part per million (vol:vol)
Pollutant loading	kg/h	

3.0   PRIMARY TREATMENT

Primary clarifiers are used to separate solids from wastewater. Circular and rectangular units typically are used. Equipment includes scraper mechanisms to move solids into a hopper, from which it is removed with sludge pumps, and skimming devices to remove floatables (scum/skimmings). Process loading criteria typically includes hydraulic and solids loading rates. See Section 1.4 of this chapter for primary sludge pumping.

TABLE 5.8   Primary clarifiers.

Parameter	SI	Comment
Detention time	h	
Particle settling rate	m/min	Alternate: m/h
Skimmings quantity (volume)	m^3/ML	
Skimmings quantity (mass)	kg/ML	
Solids collector velocity (tip speed)	m/min	
Solids concentration	%	Alternate: mg/L
Solids loading	$kg/(m^2 \cdot d)$	
Surface overflow rate (upflow velocity)	$m^3/(m^2 \cdot d)$	Same as m/d
Volume settled solids	m^3/ML	
Weir loading rate	ML/(m·d)	Alternate: $m^3/(m \cdot d)$

4.0   SECONDARY TREATMENT

4.1   Trickling Filters

Trickling filters typically are used for BOD removal but can also be designed for nitrification. Biofilm grows on the surface of a fixed media that is typically plastic with a porous structure and high surface area per unit volume to provide surface for biofilm growth. Treatment is achieved by distributing the wastewater across the surface of the media. Air is ventilated through the media to maintain aerobic conditions in the bed. Trickling filters are sometimes used with an activated sludge or solids contact process.

TABLE 5.9   Trickling filters.

Parameter	SI	Comment
Hydraulic loading rate	$m^3/(m^2 \cdot d)$	Hydraulic loading per cross-sectional area of filter; specify with or without recycle
Media loading rate	$kg/(m^2 \cdot d)$	Substrate loading per media area; specify BOD, COD, nitrogen, or other
Media size (rock)	mm	
Media specific area	m^2/m^3	
Organic loading rate	$kg/(m^3 \cdot d)$	Mass loading rate; specify BOD, COD, N, or other
Recirculation rate	flow %	Percentage based on flow:flow
Ventilation airflow	Nm^3/h	
Ventilation rate	$Nm^3/(m^2 \cdot h)$	Airflow expressed in normal m^3
Wetting rate	$ML/(m^2 \cdot d)$	Hydraulic loading rate

4.2   Activated Sludge

Activated sludge is a suspended growth biological treatment process. It achieves high microorganism (biomass) concentrations through the recycle of biologically active sludge. There are many variations of the basic biological processes to achieve nitrification, denitrification, and phosphorus removal. The liquid environment can be aerobic, anoxic, or anaerobic, or mixed. Oxygen is supplied to aerobic zones and anoxic and anaerobic zones are mixed to keep the solids in suspension.

TABLE 5.10   Activated sludge.

Parameter	SI	Comment
Aerator mixing energy	kW	Mechanical aeration systems
Blower capacity	Nm^3/h	Airflow expressed in normal m^3
Blower power	kW	
Blower pressure	kPa	
Dissolved oxygen	mg/L	
Food-to-microorganism ratio	kg/(kg·d)	Specify BOD, COD, N, or other and solids measure—volatile suspended solids (VSS), total suspended solids (TSS), etc.
Mixed liquor suspended solids	mg/L	Specify VSS or TSS basis
Mass-transfer coefficient, K_La	h^{-1}	
Mean cell residence time	d	
Mixing intensity (mechanical)	kW/ML	Mechanical aeration systems; alternative: W/m^3
Mixing intensity (air mix)	Nm^3/m^3	Air volume/tank volume, coarse bubble diffused aeration. Airflow expressed in m^3.
Mixing intensity (air mix)	Nm^3/m^2	Air volume/tank area, fine-pore diffused aeration. Airflow expressed in normal m^3.
Organic loading rate	kg/(m^3·d)	Volumetric loading rate
Oxygen supply		
Oxygen uptake rate (OUR)	mg/(L·h)	
Specific oxygen uptake rate (SOUR)	mg/(g·h)	Specify VSS or TSS basis
Field oxygen-transfer rate (OTR$_f$)	kg/h	
Field oxygen-transfer efficiency (OTE$_f$)	mass %	
Aeration efficiency (AE)	kg/kW·h	
Standard oxygen-transfer rate (SOTR)	kg/h	

(continued)

TABLE 5.10   Activated sludge (*continued*).

Parameter	SI	Comment
Standard oxygen-transfer efficiency (SOTE)	mass %	Oxygen transferred/oxygen supplied (w/w)
Standard aeration efficiency (SAE)	kg/kW·h	
Return activated sludge solids concentration	mass %	Alternate: mg/L
Settleable solids	mL/L	
Sludge volume index (SVI)	mL/g	
Sludge age	d	
Solids retention time (SRT)	d	
Waste activated sludge solids concentration	mass %	Alternate: mg/L

4.3   Integrated Biofilm Processes

The integrated biofilm process combines fixed and suspended biomass to provide secondary treatment. A porous media is added to the aeration basin to provide a surface area in the activated-sludge process to sustain fixed biological growth and increase sludge age. Integrated fixed-film activated sludge (IFAS) and moving-bed biofilm reactors (MBBR) are two common styles that are used. The IFAS process adds either fixed or suspended media to the activated-sludge bioreactor. This process includes using recycle from the secondary clarifier similar to the activated sludge to maintain a suspended biomass population in the system.

The MBBR also uses media in the aeration basin but does not maintain suspended growth in the bioreactor. It relies completely on biomass grown on the media surface. Aerated or unaerated zones are used for BOD removal and nitrification/denitrification. Many other process configurations exist. Parameters relating to the fixed film portion of IFAS and MBBR systems are shown below. Section 4.2 of this chapter provides parameters relating to the suspended-growth portion of the activated sludge system.

TABLE 5.11   Integrated fixed-film activated sludge and moving-bed biofilm reactors.

Parameter	SI	Comment
Media loading rate	kg/(m²·d)	Substrate loading per media area; specify BOD, COD, nitrogen, or other
Media specific area	m²/m³	
Media fill	volume %	Fraction based on volume/volume

(continued)

TABLE 5.11 Integrated fixed-film activated sludge and moving-bed biofilm reactors (*continued*).

Parameter	SI	Comment
Media growth (fixed film) equivalent	mg/L	Expressed per reactor volume; specify VSS or TSS basis
Media growth thickness	mm	
Media growth (fixed film)	kg/m^2	Growth on media surface; specify VSS or TSS basis
Scour aeration	$Nm^3/(m^2 \cdot h)$	Airflow expressed in normal m^3

4.4 Biological Aerated Filters

The biological aerated filter (BAF) process is a fixed biofilm process. It can be used for BOD removal, nitrification, and denitrification. This process typically includes a reactor filled with a filter media that is either in suspension or supported by a gravel layer at the foot of the filter. The dual purpose of this media is to support attached biomass and to filter suspended solids. The BAF is operated either in upflow or downflow configuration depending on manufacturer specifications. Aeration is provided to sustain the aerobic biological processes.

TABLE 5.12 Biological aerated filters.

Parameter	SI	Comment
Hydraulic loading rate	$m^3/(m^2 \cdot d)$	Alternate: $ML/(m^2 \cdot d)$
Media loading rate	$kg/(m^2 \cdot d)$	Substrate loading per media area; specify BOD, COD, nitrogen, or other. Mass loading per unit of media area.
Media specific area	m^2/m^3	
Organic loading rate	$kg/(m^3 \cdot d)$	Mass loading rate per unit of reactor volume; specify BOD, COD, N, or other
Recirculation rate	flow %	Based on feed flow to unit
Scour aeration	$Nm^3/(m^2 \cdot h)$	Airflow expressed in normal m^3
Ventilation rate	$Nm^3/(m^2 \cdot h)$	Airflow expressed in normal m^3

4.5 Membrane Bioreactors

A membrane bioreactor (MBR) combines the biological treatment step of the activated sludge process and solids separation in one reactor. Solids separation is achieved using membranes rather than secondary clarifiers. Process loading criteria typically can be divided into biological and membrane categories. Membrane bioreactors are a variation of the activated sludge process, which means the biological criteria are similar. Refer to Section 4.2 of this chapter for biological criteria.

TABLE 5.13   Membrane bioreactors.

Parameter	SI	Comment
Air scouring/membrane surface area	$Nm^3/(m^2 \cdot h)$	Airflow expressed in normal m^3
Flux	$L/(m^2 \cdot h)$	
Mixed liquor recirculation rate	flow %	Based on influent flow
Mixing energy	kW/ML	
Mixed liquor suspended solids in membrane tanks	mg/L	
Partitioning coefficient, K	unitless	
Permeability, P	$L/(m^2 \cdot kPa \cdot h)$	Flux per unit pressure
Solids loading on membrane surface	$kg/(m^2 \cdot d)$	
Specific energy	$kW \cdot h/m^3$	
Total membrane surface area	m^2	
Transmembrane pressure	kPa	

4.6   Secondary Clarifiers

Secondary clarifiers separate suspended solids from liquid and return biomass in the suspended solids to the activated sludge processes to maintain the bacteria population. As noted above, clarifiers are not required for MBBR or MBR secondary treatment systems. Circular and rectangular units are typical. Equipment includes scraper mechanisms to move solids into a hopper, skimmers to move scum to one or more scum boxes, and solids return pumps. Process loading criteria typically include hydraulic and solids loading rates.

TABLE 5.14   Secondary clarifiers.

Parameter	SI	Comment
Particle settling rate	m/h	
Skimmings quantity (volume)	m^3/ML	
Skimmings quantity (mass)	kg/ML	
Solids collector velocity (tip speed)	m/min	
Solids loading	$kg/(m^2 \cdot d)$	
Surface overflow rate (upflow velocity)	$m^3/(m^2 \cdot d)$	Same as m/d
Volume settled solids	m^3/ML	
Weir loading rate	$m^3/(m \cdot d)$	Alternative: $ML/(m \cdot d)$

5.0 ADVANCED WASTEWATER TREATMENT

Advanced wastewater treatment includes the processes of chemical addition, chemical precipitation, mixing, coagulation/flocculation, chemical adsorption, ion exchange, sedimentation, advanced filtration, gas stripping, advanced oxidation, and distillation. Standard design and operating units are provided in the tables below. Standard units for each treatment process are tabulated separately. Standard units for the properties of water are provided first as general background information.

5.1 Chemical Addition

This process involves adding chemicals to water or wastewater to cause some predetermined chemical reaction, such as coagulation or precipitation.

TABLE 5.15 Chemical addition.

Parameter	SI	Comment
Chemical storage		
Chemical storage period	d	
Dose	mg/L	Concentration of chemical applied to stream
Chemical feed		
Purity (v/v)	volume %	Active chemical available in product (volume/volume)
Purity (w/w)	mass %	Active chemical available in product (mass/mass)
Stock active chemical available	kg/m^3	Specify active chemical
Stock bulk density	kg/m^3	Specify active chemical
Stock concentration	kg/L	Specify active chemical
Stock density	kg/L	Specify active chemical
Stock feed rate	kg/h	Mass of chemical supplied to stream
Tank volume (dry)	m^3	
Tank volume (liquid)	L	Alternative: m^3

5.2 Chemical Precipitation

This process involves adding one chemical to cause precipitation of another chemical. The concentration of the chemical to be removed typically is described in equivalent units of the main chemical. For example, orthophosphate in water is typically measured in terms of equivalent milligrams per liter of phosphorus for ease of calculating required dosages of precipitants.

TABLE 5.16   Chemical precipitation.

Parameter	SI	Comment
Alkalinity	mg/L	Expressed as mg/L $CaCO_3$ equivalents
Concentration of chemicals	mg/L	Specify active chemical
Equivalent concentration of chemicals	meq/L	
Equivalent weight	g/eq	
Molarity, M	mol/L	M = [(molecular weight)(10^3 mg/g)]/ (mg/L)
Molecular weight	g/mol	
Normality, N	eq/L	N = Mn where n = equivalents/mole

5.3   Rapid Mixing

Once chemicals are added, some form of mixing typically is required to rapidly consolidate the new chemicals into the existing wastewater flow.

TABLE 5.17   Rapid mixing.

Parameter	SI	Comment
Detention time	min	Alternative: s
Mixer energy	kW	
Mixing energy	kW/ML	Alternative: W/m^3
Velocity gradient	s^{-1}	

5.4   Coagulation/Flocculation

The intent of chemical addition typically is to cause dissolved solids to separate from solution as fine suspended particles, followed by flocculation of those particles into large enough particles to settle in a clarifier.

TABLE 5.18   Coagulation/flocculation.

Parameter	SI	Comment
Basin volume	L	Alternative: m^3
Detention time, t_d	min	Alternative: h
Equilibrium constant, k	unitless	
Flocculation, Gt	unitless	Velocity gradient multiplied by time
Hydraulic retention time	min	
Mixer energy	kW	

(continued)

TABLE 5.18   Coagulation/flocculation (*continued*).

Parameter	SI	Comment
Mixing energy	kW/ML	Alternative: W/m^3
Mixing time, t	s	Alternative: min
Nernst potential, ψ_n	mV	
Potential energy of particles, ψ_o	mV	
Salt solubility constant	mol/L	pH dependent
Stern potential of particles, ψ_s	mV	ψ_o to ζ
Velocity gradient	s^{-1}	
Zeta potential of particles, ζ	mV	$= \psi_n - \psi_s$

5.5   Chemical Adsorption

Chemical adsorption involves using one chemical or compound, typically a solid, to adsorb a second chemical or compound, thereby removing the second compound from solution or suspension and trapping it on the adsorbent compound. Chemicals and compounds that can be removed by this process include suspended and colloidal particles and dissolved organic and inorganic compounds, including volatile organic compounds. The adsorbent typically is activated carbon, either granular or powdered, or "greensand"-type compounds. (Note that an "adsorbate" is the chemical being adsorbed; an "adsorbent" is the chemical or compound to which the adsorbate is being adsorbed.)

TABLE 5.19   Chemical adsorption.

Parameter	SI	Comment
Adsorbent phase concentration after absorbance	mg/g	(mg adsorbate/g adsorbent)
Ash content	mass %	Dry-weight basis
Bulk dry density of adsorbent, ρ	kg/m^3	
Concentration of adsorbate	mg/L	
Mass of adsorbate	g	Specify compound
Mass of adsorbent	g	Specify adsorbent
Mean pore radius	Å	
Particle density (in water)	kg/L	
Particle diameter	mm	
Particle size	mm	
Pore size of adsorbent	nm	
Total surface area of adsorbent	m^2/g	
Volume of reactor	m^3	Alternative: L

TABLE 5.20 Chemical adsorption with activated carbon.

Parameter	SI	Comment
Approach velocity, V_f	m/h	
Bed volume ratio, BV	m^3/m^3	Volume of carbon/volume of reactor
Bed volume, V_b	m^3	Volume of carbon in reactor; alternate: L
Breakthrough adsorption capacity, $(x/m)_b$	g/g	(g adsorbate/g adsorbent)
Breakthrough organic concentration, C_b	g/m^3	
Carbon use rate	g/m^3	
Cross-sectional area of bed or reactor, A_b	m^2	
(Effective) contact time, t	min	
Empty bed contact time, t	min	
Flowrate through granular activated carbon (GAC) column	m^3/d	
Freundlich intensity parameter, $1/n$	unitless	
Freundlich capacity factor, K_f	(mg/g) $(L/mg)^{1/n}$	(mg adsorbate/g activated carbon) (L of water/mg adsorbate)$^{1/n}$
Influent organic concentration to GAC column, C_o	mg/L	Alternative: g/m^3
Length of reactor vessel, D	m	
Mass of adsorbate adsorbed in a GAC column at breakthrough, x_b	kg	
Mass of adsorbate adsorbed per mass of adsorbent, x/m	mg/g	mg adsorbate/g adsorbent
Mass of GAC in the column at breakthrough, m_{GAC}	kg	
Operation time, t	d	
Pressure in column	kPa	
Specific throughput, V_{sp}	ML/kg	ML of contaminated water/g of GAC
Throughput volume, V_L	ML	
Time to breakthrough, t_b	d	
Void fraction, α	m^3/m^3	Volume of voids/volume of reactor

5.6   Ion Exchange

In the ion-exchange process, ions of one chemical or compound are exchanged for an ion of a different, typically less harmful chemical or compound in a solution.

TABLE 5.21   Ion exchange.

Parameter	SI	Comment
Bed area	m^2	
Bed depth	m	Dry-media basis
Bed volume	m^3	Dry-media basis
Concentration of ions	meq/L	Milliequivalents per liter
Exchange capacity of a resin	g/m^3	g/m^3 of resin; specify compound
Exchange capacity of a resin	eq/m^3	gram equivalents per cubic meter
Loading rates to reactors	$L/(m^2 \cdot min)$	
Selectivity coefficient	unitless	

5.7   Sedimentation

Sedimentation involves the precipitation of suspended solids through a quiescent water column to remove the suspended solids from the media. See Section 4.6 in this chapter for standard units of sedimentation.

5.8   High-Rate Clarification

High-rate clarification is fundamentally a rapid sedimentation process. See Section 4.6 for standard units of high-rate clarification.

5.9   Advanced Filtration

Although filtration is a common operation in water treatment, it is not typically found in conventional wastewater treatment. Filtration is, however, becoming more prevalent particularly in water reclamation applications. The lower discharge concentrations typically require additional chemical coagulation, flocculation, and precipitation or sedimentation, followed by filtration to remove the fine pin floc and suspended colloidal particles remaining after sedimentation. There are several variations of filters in use including mono-medium using sand, anthracite, or synthetic fibers and dual-medium using both sand and anthracite. They also include conventional filters, deep bed filters, continuous backwash filters and traveling bridge filters using any of the cited media. The principal differences lie in the nature of the filter media, flowrates through the filter, backwash cycles required, and the direction of flow of dirty water through the filter. Many of the filter types in use today are proprietary designs, but all use the same general parameters for design and operation.

TABLE 5.22   Advanced filtration.

Parameter	SI	Comment
Backwash rate	m/h	
Backwash duration	min	
Bed depth, D or L	m	
Bed surface area, A	m^2	
Bed volume, V	m^3	
Concentration of particles, or suspended solids, C	mg/L	
Effective size of filter media, d_{10}	mm	The diameter of the particles, measured with standard sieves at which 10% of the particles are finer
d_{60}	mm	The diameter of the particles, measured with standard sieves at which 60% of the particles are finer
Filtration rate	m/h	
Flux rate (membrane filters)	$L/(m^2{\cdot}h)$	
Hydraulic loading rate (granular filters)	$L/m^2{\cdot}min$	
Nutrient loading rate	$kg/m^3{\cdot}d$	
Solids loading rate	$kg/(m^3{\cdot}d)$	
Transmembrane pressure (membrane filters)	kPa	
Uniformity coefficient, d_{60}/d_{10}	unitless	The ratio of the particle diameter, in mm, measured with standard sieves at which 60% of the particles are finer, divided by the effective size of the particles

Head loss equation terms

Acceleration from gravity, g	$9.81\ m/s^2$	
Approach velocity, v_s	m/s	
Area of filter media, A_P	m^2	
Coefficient of compaction, C	unitless	
Coefficient of drag, C_d	unitless	
Density of media, ρ_s	kg/m^3	
Diameter of filter media particle, d	mm	
Friction factor, f	unitless	

(continued)

TABLE 5.22 Advanced filtration (*continued*).

Parameter	SI	Comment
Geometric mean diameter between sieve sizes d_1 and d_2, in mm	mm	$d_g = (d_1 d_2)^{0.5}$
Headloss, h	m	
Mass fraction of particles within adjacent sieve sizes, p	mass %	
Particle shape factor, φ	unitless	
Porosity of media, α	volume %	
Specific gravity of media	unitless	
Temperature	°C	°C = 5/9 (°F − 32)
Viscosity of fluid, μ	(N·s)/m²	
Volume of filter media, v_p	m³	
Backwash equation terms		
Air scour rate	Nm³/(m²·h)	Alternative: L/(m²·s)
Backwash rate	L/min	
Backwash volume required	volume %	Volume of backwash water/ volume of water treated
Bed expansion	m	
Density of water	kg/m³	
Expanded bed depth	m	
Upflow velocity of backwash water	m/h	
Terminal headloss	m	

5.10 Membrane Filtration

Membrane filtration is designed to separate very fine suspended and colloidal particles from water or wastewater. The term includes microfiltration, ultrafiltration, nanofiltration, reverse osmosis, and electrodialysis.

TABLE 5.23 Membrane filtration.

Parameter	SI	Comment
Average imposed pressure Gradient, ΔP_a	kPa	$= [(P_f - P_c)/2] - P_p$
Average osmotic pressure Gradient, $\Delta \Pi$	kPa	$= [(\Pi_f + \Pi_c)/2] - \Pi_p$
Average solute concentration Gradient, ΔC_i	mg/L	$= [(C_f + C_c)/2] - C_p$
Concentrate concentration, C_c	mg/L	

(continued)

TABLE 5.23   Membrane filtration (*continued*).

Parameter	SI	Comment
Concentrate flowrate, Q_c	kg/h	
Concentrate osmotic pressure, Π_c	kPa	
Concentrate pressure, P_c	kPa	
Feedwater concentration, C_f	mg/L	
Feedwater flowrate, Q_f	kg/h	
Feedwater osmotic pressure, Π_f	kPa	
Feedwater pressure, P_f	kPa	
Mass-transfer constant, solute, k_j	m/s	
Mass-transfer constant, water, k_w	m/s	
Permeate concentration, C_p	mg/L	
Permeate flowrate, Q_p	m³/h	
Permeate osmotic pressure, Π_p	kPa	
Permeate pressure	kPa	
Target particle size	μm	micron
Total membrane surface area, A	m²	
Transmembrane pressure gradient, P_m	kPa	
Transmembrane water flux rate, F_w	L/(m²·h)	

5.11   Gas Stripping

Gas stripping involves the introduction of air, or occasionally other gases, to cause the volatilization and off-gassing of volatile organic compounds, ammonia, carbon dioxide, hydrogen sulfide, and other volatile constituents in water or wastewater. The process transfers gases from a liquid phase to a vapor phase, and then accelerates the vaporization of the gases to the atmosphere.

TABLE 5.24   Gas stripping.

Parameter	SI	Comment
Change of concentration desired with time, $\partial C / \partial t$	g/(m³·s)	
Concentration in liquid bulk phase at time t, C_b	g/m³	
Concentration in liquid in equilibrium with gas, C_s	g/m³	
Concentration of constituent, C	g/m³	
Concentration of solute at any point within the tower, y	mol/mol	mol of solute/mol of solute-free gas

(*continued*)

TABLE 5.24   Gas stripping (*continued*).

Parameter	SI	Comment
Concentration of solute in gas entering the bottom of the tower, y_o	mol/mol	mol of solute/mol of solute-free gas
Concentration of solute in gas leaving the top of the stripping tower, y_e	mol/mol	mol of solute/mol of air
Concentration of solute in liquid at any point within the tower, C	mol/mol	mol of solute/mol of liquid
Concentration of solute in liquid entering the stripping tower, C_o	mol/mol	mol of solute/mol of liquid
Concentration of solute in liquid leaving the bottom of the tower, C_e	mol/mol	mol of solute/mol of liquid
Concentration of solute in liquid that is in equilibrium with the gas leaving the tower, C_o'	mol/mol	mol of solute/mol of liquid
Design of tower height		
Differential height, Δz	m	
Differential volume, ΔV	Nm^3	
Diffusion coefficient of oxygen in water, D_{O2}	cm^2/s	
Diffusion coefficient of volatile organic carbon in water, D_{VOC}	cm^2/s	
Height of transfer unit	m	
Henry's law constant	(atm·mol/ mol)/ (mol/mol)	Alternative: atm (mol of gas/mol of air)/(mol of gas/mol of water)
Mass of incoming gas per unit of time, G	mol/s	Alternative: mol of gas/s or min
Mass of incoming liquid per unit of time, L	mol/s	Alternative: mol of liquid/s or min
Mass-transfer rate per unit of volume per unit of time, r_v	$g/(m^3 \cdot s)$	
Number of transfer units	unitless	
System mass-transfer coefficient, $K_L a_{VOC}$	L/h	
System oxygen mass-transfer coefficient, $K_L a_{O2}$	L/h	
Total pressure, P_T	atm	
Tower coefficient, n	unitless	
Volumetric flowrate of liquid	m^3/s	
Volumetric mass-transfer coefficient, $K_L a$	L/s	

(continued)

TABLE 5.24   Gas stripping (*continued*).

Parameter	SI	Comment
Design parameters		
Air-to-liquid ratio, G/L	unitless	
Allowable air pressure drop, ΔP	$(N \cdot m^2)/m$	
Density of gas, ρ_G	kg/m^3	
Factor of safety, SF	% Z (height of the media in the column) or % D (diameter of the column)	
Gas loading rate, G'	$kg/(m^2 \cdot s)$	
Height-to-diameter ratio, Z/D	unitless	
Liquid loading rate, L'	$L/(m^2 \cdot min)$	Alternative: $kg/(m^2 \cdot s)$
Packing depth, Z	m	
Packing factor, C_f	unitless	
Stripping factor, S	unitless	
Viscosity of liquid, μ_L	$(N \cdot s)/m^2$	

5.12   Advanced Oxidation

Advanced oxidation typically is used to oxidize persistent organic compounds that are difficult to degrade biologically in a conventional wastewater treatment plant.

TABLE 5.25   Advanced oxidation.

Parameter	SI	Comment
Concentration of contaminants	mg/L	
Concentration of oxidant	mg/L	
Concentration, final, C_f	ng/L	
Concentration, initial, C_i	ng/L	
Electrical energy input, EE_i	kWh	
Electrical energy required per log	$kWh/(m^3 \cdot log)$	
Reduction of contaminant, EE/O	order of reduction	
Electrochemical oxidation potential, EOP	V	volts
UV light wavelength	nm	
Volume of liquid treated, V	m^3	

5.13   Distillation

The distillation process is used to separate liquids through volatilization and subsequent condensation of one of the liquids.

TABLE 5.26   Distillation.

Parameter	SI	Comment
Energy required to achieve latent heat of vaporization	kJ/kg	
Latent heat of vaporization	kJ/kg	
Temperature, T	°C	°C = 5/9 (°F − 32)

5.14   Natural Treatment Systems

Natural treatment systems are used in lieu of conventional activated sludge plants or as advanced treatment systems for secondary effluent. This approach includes various types of lagoon systems, land treatment of effluents through infiltration, and both natural and constructed wetlands. This section also includes other water reclamation and reuse options.

5.14.1   Lagoons

Lagoon systems covered by this section include stabilization lagoons (aerated, aerobic, anaerobic, and facultative) and floating aquatic plant lagoons.

TABLE 5.27   Lagoon systems.

Parameter	SI	Comment
Biochemical oxygen demand removal (BOD)	mass %	
Depth	m	
Detention time	d	
Flow-through velocity	m/s	
Hydraulic loading rate	$m^3/(ha \cdot d)$	
Organic loading rate	kg BOD/(ha·d)	
Slope	ratio	Horizontal distance to vertical distance
Surface area	ha	Alternative: m^2
Temperature	°C	
Turbidity	NTU	NTU = nephelometric turbidity units
Turbidity factor, TSS_f	(mg/L TSS)/NTU	Factor to convert TSS to NTUs

5.14.2 Land Infiltration

In land treatment, wastewater is discharged to the surface of the land where it is treated as it percolates through the soil. Three types of land treatment are recognized: slow-rate infiltration, high-rate infiltration, and overland flow. This section addresses all three types.

TABLE 5.28   Land infiltration/treatment systems.

Parameter	SI	Comment
Application period	d	Biochemical oxygen demand
Concentration of organics	mg/L	
Drying period	d	
Hydraulic loading rate	$m^3/(ha\cdot d)$	
Metals loading rate	$kg/(ha\cdot d)$	
Natural percolation rate	mm/d	
Nutrient loading rate	$kg/(ha\cdot d)$	
Organic loading rate	$kg\ BOD/(ha\cdot d)$	
Temperature	°C	

5.14.3 Wetlands

Wetlands used for treating wastewater can be either natural or constructed; both are included in the following table. The units are the same for both types of system.

TABLE 5.29   Wetland treatment systems.

Parameter	SI	Comment
Berm-side slope dimensions	ratio	Horizontal distance/vertical distance
Clay liner thickness	m	
Concentration of constituents	mg/L	
Design flowrate	m^3/d	
Distribution box spacing	m	
Gravel size	mm	
Hydraulic loading rate	$m^3/(m^2\cdot d)$	Wastewater flow applied continuously to head of subsurface-flow bed supporting emergent vegetation or to head of channel containing emergent aquatic vegetation
Hydraulic retention time	d	
Organic loading rate	$kg\ BOD/(ha\cdot d)$	Alternative: $kg\ BOD/(m^2\cdot d)$

(continued)

TABLE 5.29   Wetland treatment systems (*continued*).

Parameter	SI	Comment
Organics concentration	mg/L	
Plant spacing	m	
Plastic liner thickness	mils	
Suspended solids concentration	mg/L	
Temperature	°C	
Water depth	m	

5.14.4 Water Reclamation Options

This section includes water reuse for agricultural irrigation, landscape irrigation, industrial recycling and reuse, groundwater recharge, and other recreational, environmental, nonpotable, and potable reuse options. Bacteria concentrations in colony forming units (CFUs) relate to membrane filtration techniques, and concentrations in most probable numbers (MPNs) relate to multiple tube fermentation techniques; these are not interchangeable.

TABLE 5.30   Water reclamation and reuse options.

Parameter	SI	Comment
Bacteria concentration	CFU/100 mL	Alternative: MPN/100 mL
Contaminant concentration	mg/L	
Crop evapotranspiration, ET_c	mm/d	
Depth of water applied at the surface, D_i	mm	
Depth of water leaching below the root zone, D_d	mm	$D_d = D_i - ET_c$
Evapotranspiration, ET	mm/d	
Flow	m³/d	
Leaching fraction, LF	unitless	$LF = D_d/D_i$
Natural percolation rate	mm/d	
Potential evapotranspiration, ET_o	mm/d	
Rainfall	mm	
Salinity (based on electrical conductivity), EC_w	dS/m	Decisiemens per meter; alternative: mmho/cm 1 dS/m = 1 mmho/cm
Salinity (based on total dissolved solids, TDS)	mg/L	TDS (mg/L) = EC (dS/m or mmho/cm) × 640 (approximately)
Storage reservoir area	ha	
Temperature	°C	

6.0   DISINFECTION

Disinfection is the process of inactivating pathogenic or other organisms. Chemical and physical processes are commonly used for disinfection.

TABLE 5.31   General disinfection.

Parameter	SI	Comment
Inactivation efficiency	count %	
Inactivation efficiency	log removal	Reduction expressed in terms of log (base 10) removal
Concentration of organisms		
Concentration of viable organism in primary source	number/L	Alternative: number/g
Fecal coliform/*E. coli*/ Enteroccoci	CFU/100 mL	Alternative: fecal coliform units (FCU)
Total coliform	MPN/100 mL	
Temperature	°C	
Turbidity	NTU	nephelometric turbidity units

6.1   Chlorine

Gaseous chlorine is commonly used for disinfection. The gas is stored onsite in pressure vessels in small cylinders or large tankers. The disinfection dose is expressed in terms of chlorine (Cl_2) equivalents.

TABLE 5.32   Gaseous chlorine disinfection.

Parameter	SI	Comment
Contact time	min	
Residual concentration × contact time, Ct	mg·min/L	
Disinfection residual	mg/L	As Cl_2
Dose	mg/L	As Cl_2
Chemical feed		
Feed rate	kg/h	Chemical supply to stream as Cl_2 Alternative: kg/d
Purity (v/v)	volume %	Active chemical available in product (volume/volume)
Purity (w/w)	mass %	Active chemical available in product (mass/mass)
Solubility concentration	mg/L	

(continued)

TABLE 5.32   Gaseous chlorine disinfection (*continued*).

Parameter	SI	Comment
Chemical storage		
Chemical storage period	d	
Gas cylinder capacity	kg	As Cl_2
Gas storage capacity	kg	As Cl_2
Tank capacity	Mg	Megagram, also tonne or metric ton
Tank volume	L	Alternative: m^3

6.2   Sodium Hypochlorite

Sodium hypochlorite is a liquid form of chlorine typically used for disinfection. Sodium hypochlorite is stored onsite in tanks and fed into the water. The most common chemical used is sodium hypochlorite (NaClO). The disinfection dose is expressed in terms of chlorine (Cl_2) equivalents.

TABLE 5.33   Sodium hypochlorite disinfection.

Parameter	SI	Comment
Contact time	min	
Dose	mg/L	
Residual concentration × contact time, Ct	mg·min/L	
Residual	mg/L	As Cl_2
Chemical storage		
Chemical storage period	d	
Tank volume	ML	Alternative: m^3
Liquid sodium hypochlorite feed		
Purity (w/w)	%	Active chemical available in product
Stock bulk density	kg/L	As Cl_2
Stock chlorine content	kg/L	As Cl_2
Stock feed rate	kg/h	As Cl_2
Stock feed rate	L/h	As liquid sodium hypochlorite

6.3   Ultraviolet

Ultraviolet disinfection relies on light in the UV range (254 nm) to inactivate pathogenic organisms. The process uses UV lamps that are placed in protective tubes and inserted in the water. The disinfection intensity typically is expressed in terms of the lamp power output.

TABLE 5.34   Ultraviolet disinfection.

Parameter	SI	Comment
Contact time	s	
Distance in the direction of irradiation	cm	
Dose	mW·s/cm^2	Alternative: mJ/cm^2
Germicidal output/input	%	Alternative: log removal (base 10)
Hydraulic loading rate	L/(lamp bank·min)	
Inactivation constant	cm^2/(mW·s)	
Lamp life	H	
Lamp operating temperature	°C	
Light intensity	mW/cm^2	
Path length in absorbing medium	cm	
Power/lamp	W	
Sleeve/ballast life	yr	
UV transmittance	transmittance %	
Vapor pressure in lamp	Pa	
Wavelength of radiation	nm	

6.4   Ozone

Ozone is generated onsite for disinfection or oxidation. Air or pure oxygen (typically in the form of liquid oxygen or generated onsite) is used as a feed to the ozone generator.

TABLE 5.35   Ozone disinfection.

Parameter	SI	Comment
Concentration of ozone	mg/L	Alternate: ppm$_v$ Based on gas flow
Contact time	min	
Dose	mg/L	Based on liquid flow
Liquid oxygen capacity	Mg/d	Megagram, also tonne or metric ton
Ozone generation system pressure	kPa	
Ozone generator capacity	kg/h	Equipment capacity
Ozone power required	kW·h/kg	
Ozone use	kg/d	Daily consumption
Residual	mg/L	In liquid

7.0   SOLIDS THICKENING

Thickening is used to significantly increase the concentration of liquid solids drawn from primary and secondary sedimentation tanks. Thickening improves the design and operations efficiencies of subsequent solids treatment processes.

7.1   Gravity

Gravity thickeners are used to thicken primary and secondary solids from wastewater processes. Circular and rectangular units are typical. Equipment includes scraper mechanisms to move solids into a hopper and solids pumps. Process loading criteria typically includes hydraulic and solids loading rates.

TABLE 5.36   Gravity thickeners.

Parameter	SI	Comment
Hydraulic overflow rate	$m^3/(m^2 \cdot d)$	
Polymer requirement	mg/kg	
Solids capture	mass %	
Solids loading rate	$kg/(m^2 \cdot d)$	
Thickened solids concentration	mass %	
Volume thickened solids	m^3/ML	
Weir loading rate	$m^3/(m \cdot d)$	

7.2   Dissolved Air Flotation

Dissolved air flotation thickeners are used to thicken primary and secondary solids from wastewater processes. Circular and rectangular units are typical. Equipment includes recycle pressurization systems, float and scraper mechanisms to move solids into a hopper, and solids pumps. Process loading criteria typically includes hydraulic and solids loading rates.

TABLE 5.37   Dissolved air flotation thickeners.

Parameter	SI	Comment
Air-to-solids ratio	kg/kg	
Recycle rate	flow %	
Solids collector velocity	m/min	
Hydraulic surface overflow rate (upflow velocity)	$m^3/(m^2 \cdot d)$	
Solids loading	$kg/(m^2 \cdot d)$	
Solids capture	mass %	
Thickened solids concentration	mass %	
Polymer requirement	mg/kg	
Volume thickened solids	m^3/ML	
Weir loading rate	$m^3/(m \cdot d)$	

7.3   Rotary Drums

Rotary-drum thickeners are used to thicken primary and secondary solids from wastewater processes. Equipment includes raw and thickened solids feed pumps, thickener units, and polymer feed equipment. Process loading criteria typically includes hydraulic and solids loading rates.

TABLE 5.38   Rotary-drum thickeners.

Parameter	SI	Comment
Drum diameter	m	
Hydraulic loading	L/(unit·min)	
Polymer requirement	mg/kg	
Rotation speed	rpm	
Screen opening size	mm	
Solids capture	%	
Solids loading	kg/(m²·d)	
Thickened solids concentration	mass %	
Volume thickened solids	m³/ML	

7.4   Gravity Belts

Gravity-belt thickeners are used to thicken primary and secondary solids from wastewater processes. Equipment includes raw and thickened sludge feed pumps, thickener units, and polymer feed equipment. Process loading criteria typically includes hydraulic and solids loading rates.

TABLE 5.39   Gravity-belt thickeners.

Parameter	SI	Comment
Belt speed	m/min	
Belt width	m	
Hydraulic loading rate	L/(m·min)	
Liquid feed rate	L/min	
Polymer requirement	kg/Mg	Alternative: g/kg or kg/Mg; dry-solids basis
Solids capture	mass %	
Solids feed rate	kg/h	
Solids loading rate	kg/(m·h)	
Thickened solids concentration	mass %	
Volume thickened solids	m³/ML	
Washwater pressure	kPa	
Washwater required	L/(m·min)	

7.5   Centrifuges

Centrifuge thickeners are used to thicken primary and secondary solids from wastewater processes. Equipment includes raw and thickened sludge feed pumps, thickener units, and polymer feed equipment. Process loading criteria typically includes hydraulic and solids loading rates.

TABLE 5.40   Centrifuge thickeners.

Parameter	SI	Comment
Hydraulic loading	L/min	
Bowl speed	rpm	
Polymer requirement	kg/Mg	Alternative: g/kg or kg/Mg; dry-solids basis
Solids capture	mass %	Determined as mass removal efficiency (mass/mass)
Solids loading	kg/h	
Thickened solids concentration	mass %	mass/mass
Volume thickened solids	m^3/ML	

8.0   STABILIZATION

Solids must be stabilized to comply with regulatory requirements before disposal. Various forms of stabilization can be used. Many methods of stabilization reduce the amount of subsequent solids generated; other methods produce solids with beneficial soil qualities. The following sections present parameters associated with some of the most common methods used for solids stabilization.

8.1   Aerobic Digestion

Often viewed as an extension of the activated sludge process, aerobic digestion stabilizes organic solids and cell matter endogenously in the presence of excess oxygen to the end products of water, carbon dioxide, and ammonia. Factors affecting the aerobic digestion process include tank volume, solids loading rate, dissolved oxygen concentration, mixing requirements, and temperature.

TABLE 5.41   Aerobic digestion.

Parameter	SI	Comment
Aeration (blower)	Nm^3/(m^3·h)	Airflow expressed in normal m^3
Aeration (mechanical)	kW/ML	Alternative: kW/m^3
Dissolved oxygen concentration	mg/L	
Hydraulic retention time	d	
Solids loading	kg of VS/(m^3·d)	VS = volatile solids
Solids retention time	d	
Volatile solids destruction	mass %	Alternative: kg/kg

8.2   Anaerobic Digestion

Anaerobic digestion is a fermentation process that converts organic solids, in the absence of oxygen, to carbon dioxide and methane. Several parameters must be considered when using anaerobic digestion for stabilization: solids loading, mixing, gas production rate, hydraulic retention time, temperature, pH, and nutrient conditions, and related chemical conditions.

TABLE 5.42   Anaerobic digestion.

Parameter	SI	Comment
Digester area	m^2	
Digester volume	ML	
Flow (liquid stream)	ML/d	
Flow (liquid stream)	kg/d	For heat calculations
Gas composition (methane, CO_2, etc.)	volume %	Gas fraction volume/volume
Gas flow	Nm^3/h	Airflow expressed in normal m^3
Gas production	$Nm^3/(kg of VS \cdot d)$	Airflow expressed in normal m^3
Gas production	Nm^3/kg	Specify volatile solids applied or destroyed; airflow expressed in normal m^3
Heat loss coefficient	$MJ/(m^2 \cdot °C \cdot h)$	
Heating	kJ/d	
Heating value of gas	kJ/Nm^3	Airflow expressed in normal m^3
Hydraulic retention time	d	
Mixer energy	kW	
Mixing energy	kW/ML	Alternative W/m^3
Solids concentration	mass %	Specify VS, total solids, etc. mass/mass basis
Solids loading	kg of $VS/(m^3 \cdot d)$	
Solids retention time	d	
Specific heat stream	$MJ/(m^3 \cdot °C)$	
Temperature	°C	
Alkalinity	mg/L	Expressed as mg/L $CaCO_3$ equivalents
Volatile acids	mg/L	
Volatile acids : alkalinity	—	Ratio
pH	s.u.	Standard unit
VS destruction	mass %	Alternative: kg/kg

8.3   Lime Stabilization

Liquid and dewatered solids can be stabilized by adding sufficient quantities of lime to elevate the pH for a predetermined amount of time. Under these conditions, high kill levels of bacteria and pathogens are achieved.

TABLE 5.43   Lime stabilization.

Parameter	SI	Comment
Detention time	d	
Lime dose	kg lime/kg TS	
pH	Standard units	
Temperature	°C	

8.4   Composting

Composting, under mostly aerobic conditions, will stabilize organic solids into humus-like material that can be used as a soil amendment or conditioner. A bulking agent is used to optimize the moisture content, and aerobic conditions are maintained by either mechanical aeration or turning.

TABLE 5.44   Composting.

Parameter	SI	Comments
Aeration rate	Nm^3/h	Airflow expressed in normal m^3
Carbon-to-nitrogen ratio	kg/kg	mass/mass basis
Composting time	d	
Percent solids	%	mass/mass basis
Screen size	mm	
Temperature	°C	
Volatile solids destruction	kg/kg	

9.0   DEWATERING

Solids dewatering reduces the amount of material that needs to be processed or handled by removing free water. Mechanical dewatering devices of various designs and drying beds typically are used. Dewatering of solids enhances the efficiency of subsequent solids handling and reduces hauling and disposal costs.

9.1   Belt Filter Presses

Belt filter presses typically dewater by passing chemically conditioned solids through a gravity drainage section followed by a pressure section where the solids are sandwiched between porous belts and passed over a series of wide rollers. Dewatered solids cake is then scraped from the belt. Water released during dewatering and belt wash is discharged back to the wastewater treatment train.

TABLE 5.45   Belt filter presses.

Parameter	SI	Comment
Belt speed	m/min	
Belt width	m	
Cake	mass %	mass/mass
Feed solids concentration	mass %	mass/mass
Feed velocity	m/s	
Hydraulic loading rate	L/(m·min)	
Liquid feed rate	L/min	
Operational frequency	h/d	
Polymer required	kg/Mg	Alternative: g/kg dry-solids basis
Dewatered solids concentration	mass %	mass/mass
Solids recovery	mass %	mass/mass
Solids feed rate	kg/h	
Solids loading rate	kg/(m·h)	
Wash water flow	L/(m·min)	
Wash water pressure	kPa	

9.2   Centrifugation

Centrifuges use dynamic forces to enhance sedimentation resulting in mechanical dewatering devices with high throughput capacity and small area requirements. Chemically conditioned liquid solids are fed on a continuous basis to a solid-bowl centrifuge and are separated into centrate and dewatered solids cake.

TABLE 5.46   Centrifugation.

Parameter	SI	Comment
Cake	mass %	mass/mass
Centrifugal acceleration	m/s^2	
Feed solids concentration	mass %	mass/mass
Liquid feed rate	L/min	
Operational frequency	h/d	
Polymer required	kg/Mg	Alternative: g/kg dry-solids basis
Radius of rotating body of liquid	m	

(continued)

TABLE 5.46   Centrifugation (*continued*).

Parameter	SI	Comment
Rotational speed of centrifuge	rpm	
Differential scroll speed	rpm	
Settling velocity	m/s	
Solids feed rate	kg/h	
Dewatered solids concentration	mass %	mass/mass
Solids recovery	mass %	mass/mass

9.3   Screw Presses

Screw presses use a cylindrical, perforated screen and a conveyor screw to continuously dewater chemically conditioned liquid solids. A gravity section is used to achieve initial dewatering of the slurry. The solids then enter the screw section of the press where increasing pressure and friction removes additional water producing a dewatered solids cake.

TABLE 5.47   Screw presses.

Parameter	SI	Comment
Cake	mass %	mass/mass
Feed solids concentration	mass %	mass/mass
Filtration pressure	kPa	
High-pressure spray	kPa	
Liquid feed rate	L/min	
Operational frequency	h/d	
Polymer required	kg/Mg	Alternative: g/kg dry-solids basis
Precoat material	kg/m^2	
Press yield	kg/(m^2·h)	
Solids feed rate	kg/h	
Dewatered solids concentration	mass %	mass/mass
Solids recovery	mass %	mass/mass

9.4   Drying Beds

Drying beds dewater solids through a combination of gravity drainage of free water and evaporation. Nonthickened, stabilized liquid solids are applied to drying beds and allowed to dry for extended periods of time. The beds are constructed entirely of sand or are partially lined with bituminous material. Both types of beds include underdrains. Dried solids are removed from the beds manually or by mechanical scraping equipment.

TABLE 5.48   Drying beds.

Parameter	SI	Comment
Area	m^2	
Cake	% dry solids	
Feed solids concentration	% dry solids	
Free water pan evaporation rate	mm/yr	
Precipitation rate	mm/yr	
Daily solids loading rate	$kg/(m^2 \cdot d)$	
Annual solids loading rate	$kg/(m^2 \cdot yr)$	
Dewatered solids concentration	mass %	mass/mass

10.0   THERMAL TREATMENT

Thermal treatment of dewatered solids can be used to convert stabilized solids into a dry, marketable product. Thermal treatment also can destroy organic solids, reduce pathogens, and significantly reduce the amount of material requiring ultimate disposal. Thermal treatment of solids includes heat drying and incineration, both of which are supported by a variety of technological approaches. Following are basic parameters related to the various forms of thermal treatment available.

10.1   Heat Drying

Evaporating water using heat dryers further reduces the moisture content of dewatered solids resulting in a product that has improved handling and marketability characteristics. Heat drying solids reduces pathogens and produces a product that has lower transportation costs.

TABLE 5.49   Heat drying.

Parameter	SI	Comment
Dried solids bulk density	kg/m^3	
Dried solids concentration	mass %	mass/mass
Dryer efficiency, η	unitless	Dryer dependent
Evaporation rate	kg/h	Kilogram of water evaporated per hour
Evaporative heat requirements	kJ/kg	Heat required per kilogram of water requirements evaporated
Feed solids concentration	mass %	mass/mass
Fuel heat value	kJ/kg	Heat value per kilogram of specific fuel
Temperature	°C	
Total heat requirements	kJ/kg	
Wet solids loading rate	$kg/(m^2 \cdot h)$	

10.2 Incineration

Incineration is used to thermally convert dewatered organic solids to water, carbon dioxide, and ash. Pathogens and toxic compounds are destroyed. Incineration significantly reduces the amount of material requiring disposal and has the potential of beneficial energy recovery. Air emissions and the disposal of ash residuals must be carefully controlled.

TABLE 5.50   Incineration.

Parameter	SI	Comment
Ash density	kg/m^3	
Ash production rate	kg/h	
Combustion air flow rate	kg/h	
Feed solids concentration	mass %	mass/mass
Flue gas flow rate	kg/h	
Flue gas scrubber discharge particulate concentration	kg/Mg	
Latent heat of evaporation	kJ/kg	
Mass (ash and flue gas constituents)	kg	
Mass (water)	kg	
Scrubber particulate removal efficiency	kg/kg	mass/mass
Solids heating value	kJ/kg	
Specific heat (ash and flue gas constituents)	$kJ/(kg\cdot°C)$	
Temperature	°C	
Total heat requirements	kJ	
Wet solids loading rate	$kg/(m^2\cdot h)$	

11.0   SOLIDS REUSE AND DISPOSAL

Treated solids must ultimately be removed from the treatment site and reused or disposed of in an environmentally acceptable manner. The basic parameters associated with hauling, land application, and landfilling of treated solids are presented below.

11.1   Land Application

Land application of stabilized solids helps to condition soil and results in beneficial reuse of nutrients. Solids can be land applied either as liquid slurries or as dewatered cake. Application rates are governed by crop uptake of nutrients. Application rates must also consider limiting factors associated with long-term accumulation of metals and other constituents contained in the solids.

TABLE 5.51 Land application.

Parameter	SI	Comment
Area	ha	
Application rate (metals basis):		
Limiting metal concentration in solids	mg/kg	
Annual limiting metal loading rate	kg/(ha·yr)	
Solids loading rate	Mg/(ha·yr)	
Application rate (nitrogen basis):		
Crop-available nitrogen in solids	g/kg	
Annual crop uptake of nitrogen	kg/(ha·yr)	
Solids loading rate	Mg/(ha·yr)	
Application rate (phosphorus basis):		
Crop-available phosphorus in solids	g/kg	
Crop-available phosphorus (previous applications)	kg/ha	
Annual crop uptake of phosphorus	kg/(ha·yr)	
Solids loading rate	Mg/(ha·yr)	
Area	ha	
Dewatered solids concentration	mass %	mass/mass

11.2 Hauling and Landfilling

Stabilized solids typically are transported by hauling either as slurries or dewatered cake. Hauling costs and landfill tipping fees are determined based on either a volumetric or mass basis.

TABLE 5.52 Hauling and landfilling.

Parameter	SI	Comment
Dewatered solids concentration	mass %	mass/mass
Dewatered solids bulk density	kg/m^3	
Dewatered solids haul measure:		
Volumetric basis	m^3	
Mass basis	Mg	Megagram, also tonne or metric ton
Liquid solids concentration	mass %	mass/mass
Landfill tipping fees		
Volumetric basis	$/m^3	
Mass basis	$/Mg	

Chapter 6

Facilities

1.0	CHEMICAL SYSTEMS	66
	1.1 Basic Units	66
	1.2 Aquatic/Solution Chemistry	67
	1.3 Reaction Rates and Equilibrium Constants	68
	1.4 Chemical Handling and Feeding	68
2.0	ODOR CONTROL AND AIR EMISSIONS	69
	2.1 Odor Characterization	69
	2.2 Odor Control Technologies	70
	2.2.1 *Liquid-Phase Treatment*	70
	2.2.2 *Biological Treatment*	70
	2.2.3 *Chemical–Physical Treatment Systems*	71

	2.2.3.1 *Gas-Scrubber Systems*	71
	2.2.3.2 *Dry Adsorption Systems*	71
	2.2.4 *Combustion and Incineration*	72
3.0	CONSTRUCTION QUANTITIES	72
	3.1 Data Required for Quantifying Construction Items	72
	3.2 Typical Construction Quantities	72
4.0	ELECTRICAL AND CONTROL SYSTEMS	73
	4.1 Energy Units	73
	4.2 Electrical Systems	73
	4.2.1 *Direct Current Units*	73

4.2.2 *Alternating*
     Current Units   73

4.3   Control Systems   74

4.3.1 *Memory*   74

4.3.2 *Data Types*   75

4.3.3 *Inputs and*
     Outputs   75

4.3.4 *Communica-*
     tions   76

5.0   HEATING,
VENTILATION, AND
AIR CONDITIONING   76

1.0   CHEMICAL SYSTEMS

1.1   Basic Units

TABLE 6.1   Quantities and properties.

Parameters	SI	Comments
Amount of substance	mol	
Molecular weight or atomic weight	g/mol	
Concentration[a]	g/L volume % mass % mass ppm volume ppm	Parts per million (ppm)
Density	kg/m^3	
Specific weight	kN/m^3	Weight per unit volume
Specific gravity	unitless	Relative density or specific weight compared to reference, typically water
Dynamic viscosity	$N{\cdot}s/m^2$	Resistance of a fluid to tangential or shear stress
Kinematic viscosity	m^2/s	Dynamic viscosity divided by density
Equivalent (eq)	C	Moles of ionic charge example: 1 mol of OH^- = −1 eq

[a]For dilute environmental aqueous solutions (where the density of the solution is approximately 1 000 000 mg/L) mg/L, mass ppm, and volume ppm are approximately equal.

1.2   Aquatic/Solution Chemistry

Many non-species-specific parameters for water quality are expressed relating to a typical or relevant compound, such as ammonia-nitrogen. In this example, only the nitrogen portion of the ammonia is identified. A 1-mol solution of ammonia (NH_3, molecular weight 17.0) contains 17 mg/L of ammonia (NH_3), including 14 mg/L of nitrogen. This is typically expressed as 14 mg/L of ammonia-nitrogen, or 14 mg/L NH_3-N. Parameters that are frequently expressed in this manner include ammonia, total nitrogen, and total Kjeldahl nitrogen (as N); phosphates and total phosphorus (as P); free, available, and residual chlorine (as Cl_2); and hardness and alkalinity (as $CaCO_3$). Hardness and alkalinity are further described later in this section.

TABLE 6.2   Solutions.

Parameters	SI	Comments
Molar	mol/L	Moles solute per liter solution
Normal	mol/L	Equivalents solute per liter solvent
Molal	mol/kg	Moles solute per kilogram solvent
Mole fraction	mol/mol	Moles solute per total moles in solution
Mole ratio	unitless	mol compound A/mol compound B
Weight ratio	unitless	kg compound A/kg compound B
pH (hydrogen ion)	—	Negative logarithm of hydrogen ion activity in aqueous solution
Hardness	mg/L as $CaCO_3$	Sum of polyvalent cations in solution
Calcium hardness	mg/L as $CaCO_3$	Hardness attributable to calcium
Magnesium hardness	mg/L as $CaCO_3$	Hardness attributable to magnesium
Activity	mol/L	Effective concentration, product of concentration and unitless activity coefficient
Alkalinity	mg/L as $CaCO_3$ to reach titration end point	Acid neutralizing capacity to reach titration end point
Acidity	mg/L as NaOH to reach titration end point	Base neutralizing capacity to reach titration end point
Ionic strength	mol/L ionic impact on solution	Measure of deviation from ideal behavior because of ionic effects

(continued)

TABLE 6.2   Solutions (*continued*).

Parameters	SI	Comments
Reaction potential	V	
Oxidation-reduction potential	mV	Electrode-specific response
Dissolved oxygen	mg/L	
Surface tension	N/m	Attraction of a liquid to itself compared to adjacent medium
Vapor pressure	kN/m^2	Pressure exerted by vapor of a liquid on surrounding atmosphere

1.3   Reaction Rates and Equilibrium Constants

Many terms in this section typically are presented in the negative logarithm form, with "p" preceding the term, similar to pH. Any term presented with the negative logarithm convention should be unitless.

TABLE 6.3   Reaction rates and constants.

Parameters	SI	Comments
Equilibrium constant, K_a	unitless (mol/L/mol/L)	Molar ratio
Activity coefficient, γ	unitless (mol/L/mol/L)	Molar ratio
Solubility constant/solubility product, K_{sp}	unitless (mol/L/mol/L)	Molar ratio
Saturation constant	unitless (mol/L/mol/L)	Molar ratio
Stability index, I	unitless (specify index)	Corrosion and scaling potential measure; specify index

1.4   Chemical Handling and Feeding

TABLE 6.4   Chemical feeding.

Parameters	SI	Comments
Active content	unitless	Mole ratio of active ingredient to total bulk chemical product
Available content	unitless	Mole ratio of stoichiometrically available ingredient (for desired purpose) to total bulk chemical product
Dose	mg/L	

(continued)

TABLE 6.4 Chemical feeding (*continued*).

Parameters	SI	Comments
Free or combined chlorine	mg/L as Cl_2	
Mixing energy	J/m^3	
Power number	unitless	
Flow number	unitless	

2.0 ODOR CONTROL AND AIR EMISSIONS

Throughout this section, gas volumes will be listed in Normalized units. For further explanation of standard gas conditions for this manual, see Chapter 2, Section 2.0.

2.1 Odor Characterization

Before a control technology can be selected and designed, an engineer must define the gas stream being treated, the odorant of concern, and the performance criteria for odor removal. The following table describes units for the initial characterization of the odor.

TABLE 6.5 Odor characterization.

Parameter	SI	Comments
Gas flow rate	Nm^3/min	
Pollutant concentration in gas	g/Nm^3	Use appropriate prefixes as necessary
Odor concentration	OU/Nm^3	
Intensity	mg/Nm^3	Parts per million volume basis
Odor threshold or odor dilution to threshold	Nm^3/Nm^3	Unitless, total volume of odor sample plus diluents divided by volume of odor sample
Flux	$Nm^3/m^2 \cdot s$	
Surface area	m^2	
Temperature	°C	
Dispersion factor	s/m^3	
Point-source emission rate	g/s	
Area emission rate	$g/m^2 \cdot s$	
Ventilation rate	AC/h	AC = air change

2.2   Odor Control Technologies

This section details units for odor treatment systems. The design and application of these systems is not discussed here but can be found in other WEF manuals such as *Design of Municipal Wastewater Treatment Plants* (MOP 8) and *Control of Odors and Emissions from Wastewater Treatment Plants* (MOP 25).

2.2.1   Liquid-Phase Treatment

The following table describes units for systems that treat odorous compounds in the liquid phase, such as air or oxygen injection, chemical oxidation, nitrate addition, addition of iron salts, and pH control.

TABLE 6.6   Liquid-phase treatment.

Parameter	SI	Comments
Dissolved gas concentration	mg/L	
Gas requirements	kg/d	
Gas pressure	kPa	
Liquid requirements	m^3/d	
Volume	Nm^3	
Air flow rate	Nm^3/s	
Liquid flow rate	m^3/s	

2.2.2   Biological Treatment

Biological treatment of odor in the gas phase using autotrophic, heterotrophic, and biological-uptake processes. The following table describes units used in the design of biological filtration systems, bioscrubbers, and biotrickling filters.

TABLE 6.7   Biological treatment.

Parameter	SI	Comments
Bed depth	m	
Empty bed contact time	h	
Residence time	h	
Surface loading rate	$m^3/(m^2 \cdot h)$	
Volumetric loading rate	$Nm^3/(m^3 \cdot h)$	
Liquid application rate	$m^3/(m^2 \cdot d)$	
Surface area	m^2	
Air flow rate	Nm^3/min	
Back pressure	kPa	
Elimination rate	$g/(Nm^3 \cdot h)$	

2.2.3 *Chemical–Physical Treatment Systems*

2.2.3.1 *Gas Scrubber Systems*

Gas-phase scrubbers are systems that absorb the odorant from the gas stream into a liquid stream or which chemically oxidize the odor causing compound. Examples include packed-bed wet scrubbers, misting-scrubber systems, multiple-stage scrubbers, and catalytic oxidation.

TABLE 6.8   Gas scrubber systems.

Parameter	SI	Comments
Empty bed gas velocity	mm/s	
Packing depth	cm	
Scrubbant feed rate	L/h	
Scrubbant recirculation rate	$g/m^2 \cdot s$	
Gas loading rate	$g/(m^2 \cdot s)$	
Makeup water flow	L/min	
Headloss through vessel	kPa	
Control parameters potential	mV	
Exhaust odorant concentration	mg/Nm^3	Parts-per-million volume basis
Mass-transfer	$kg/m^2 \cdot h$	
Temperature	°C	
Height of Transfer Units (HTU)	kPa	Measure of packing separation effectiveness
Mesh	mm	
Drop size	µm	

2.2.3.2 *Dry Adsorption Systems*

In these systems, the odor particles from the gas stream are trapped on the external or internal surface of solid particles. Activated carbon systems and iron sponges are the most common systems that apply to these systems.

TABLE 6.9   Dry adsorption systems.

Parameter	SI	Comments
Gas velocity	m/min	
Packing depth	m	
Residence time	s	
Pressure drop	kPa	
Carbon surface area	ha or m^2	
Airflow	Nm^3/min	
Adsorption capacity	g/cm^3 or g/g	
Carbon life	days or weeks	

2.2.4   *Combustion and Incineration*

Combustion and incineration also will remove odor particles from the gas stream. This can be achieved with a flare, regenerative or recuperative thermal oxidizers, and catalytic oxidizers.

TABLE 6.10   Combustion and incineration.

Parameter	SI	Comments
Gas velocity	m/min	
Concentration	mg/Nm^3	
Flue-gas flow	Nm^3/h	
Temperature	°C	
Retention time	s	
Pressure drop	kPa	

3.0   CONSTRUCTION QUANTITIES

3.1   Data Required for Quantifying Construction Items

To prepare a detailed estimate of construction quantities, the following data are required:

- Plans, sections, and other relevant details of the work.

- Specifications indicating the exact nature and class of materials to be used.

To properly quantify construction items, the drawings must be clear, true to fact and scale, complete, and fully dimensioned. The selected unit of measure should be simple and convenient to measure, record, and understand, and should provide fair payment for the work involved. The result of the selected unit of measure should yield quantities neither too small nor too large and that can be accurately field measured.

3.2   Typical Construction Quantities

TABLE 6.11   Typical construction quantities.

Item Type	SI
Earthwork (cut/fill)	m^3
Underground piping (sewer, water main, etc.)	m
Piping structures (manholes, wells, etc.)	each
Concrete pavement	m^2
Asphalt pavement	Mg
Concrete sidewalk	m^2
Clearing/grubbing	ha
Traffic control	lump sum, LS

4.0   ELECTRICAL AND CONTROL SYSTEMS

4.1   Energy Units

Energy is the ability of a physical system to perform work. Joules is a measurement of energy.

TABLE 6.12   Energy parameters and units.

Parameter	SI	Comments
Energy	joule (J)	

4.2   Electrical Systems

Electrical distribution systems are found within water or wastewater facilities, such as those electrical systems found in pump stations. This includes the power-carrying circuitry and the associated control wiring. Typically, the electrical systems can be divided into two types of currents: direct current (DC) and alternating current (AC).

4.2.1   Direct Current Units

TABLE 6.13   Direct current.

Parameter	SI	Comments
Potential	volt (V)	VDC (for Volts DC) also is used
Current	amp (A)	ADC (for Amps DC) also is used
Resistance	ohm (Ω)	
Conductance	Siemens (S)	
Capacitance	farad (F)	
Inductance	henry (L)	
Power	watt (W)	

4.2.2   Alternating Current Units

TABLE 6.14   Alternating current.

Parameter	SI	Comments
Potential	volt (V)	VAC (for Volts AC) is also used
Current	amp (A)	AAC (for Amps AC) is also used
Resistance	ohm (Ω)	
Conductance	Siemens (S)	
Capacitance	farad (F)	
Inductance	henry (L)	
Real power	watt (W)	

(continued)

TABLE 6.14 Alternating current (*continued*).

Parameter	SI	Comments
Apparent power	VA	
Reactive power	VAR	
Watt hour	Wh	Used by power companies to bill for power use. Typically measured in kilowatt-hour (kWh).
Reactance	ohm (Ω)	
Impedance	ohm (Ω)	
Frequency	hertz	
Phase angle	degrees	
Harmonics	unitless	Measured in multiples of the base frequency
Short-circuit current rating	amp (A)	Also "available fault current"; typically indicated in thousands of amps (KA)

4.3 Control Systems

Control systems have evolved significantly in the past 50 years. Traditional control systems used discrete devices for each function in the control scheme; units for these are described in Table 6.13 and Table 6.14. Advanced control uses microprocessor-based devices that can consolidate many control functions into a single device. As control has evolved, new terms have developed that are unique to this area of engineering. Control systems connect to instrumentation and other field devices. Although the signals from the instruments are somewhat standardized, the parameters are unique to the type of instrument used.

4.3.1 Memory

TABLE 6.15 Automation system memory.

Parameter	Unit	Comments
Bit	bit	A memory location that can either be on or off
Byte	byte	Eight consecutive bits
Word	word	An area of consecutive bits that represents a single piece of data
Tag	tag	A piece of memory that represents a single parameter; tag can be a bit, byte, or word

4.3.2 Data Types

TABLE 6.16 Automation system data types.

Parameter	Unit	Comments
Decimal	word or tag	Base-10 numbering system that uses 0–9
Binary, digital, or Boolean	bit or tag	Base-2 numbering system that uses 1 or 0
Hexadecimal	word or tag	Base-16 number system that uses numbers 0–9 and letters A–F
Octal	word or tag	Numbering system that uses 0–7
Binary-coded decimal	word or tag	Decimal-based data type where each digit in a decimal number is converted to a binary number
Floating point/real	word or tag	Decimal-based data type that allows for the use of a decimal
Word/integer	word or tag	Decimal-based data type that allows only for whole numbers
String	tag	Data type that allows for the use of text
Engineering units		Term used to express data that have been converted to reflect the devices measurement unit

4.3.3 Inputs and Outputs

Inputs and outputs (I/O) are connections to real-world devices. This term refers only to devices that were hardwired to a control. With networks that reach the device level, however, multiple points of I/O can be accomplished over a single physical connection. Although there are different types of I/O, only three are commonly found in water and wastewater treatment facilities.

TABLE 6.17 Input and output types.

Parameter	Unit	Comments
Digital	bit	Represents on-or-off condition
Analog	0–10VDC, 0–20mADC, or 4–20mADC	Represents a single point of variable data; typically a voltage of current range is used to represent the range of data
Load dells	mVDC	An analog variant that uses a millivolt DC range to represent weight

4.3.4   *Communications*

TABLE 6.18   Communications speed measurement systems.

Parameter	Unit	Comments
Gross bit rate	bit/s	Rate of transmission
Baud rate	bd	Symbols, or pulses, per second; a symbol may have multiple bits
Throughput	bit/s	Rate of successful message delivery

5.0   HEATING, VENTILATION, AND AIR CONDITIONING

TABLE 6.19   Heating, ventilation, and air conditioning parameters and units.

Parameters	Unit	Comments
Energy	kJ	kilojoule
Power	kW	kilowatt
Heat	kJ	kilojoule
Heat flow	kJ/h	kilojoule per hour
Airflow	Nm^3/s	Normal cubic meter per second
Velocity	m/s	meter per second
Temperature	°C	degrees Celsius
Fuel value	kJ/m^3	kilojoule per cubic meter
Sound pressure level	dB	decibel
Frequency	Hz	hertz
Pressure	kPa	kilopascal
Liquid volume	L	liter
Gas volume	Nm^3	Normal cubic meter
Heat loss	kJ/m^2	kilojoule per square meter

Chapter 7

Water Reclamation and Reuse

1.0 RECLAIM, RECYCLE,   2.0 UNIQUE
     OR REUSE   77   PARAMETERS   78

1.0 RECLAIM, RECYCLE, OR REUSE

Reclaimed and recycled water is a new water supply derived from wastewater and, sometimes, stormwater. Reclaimed and recycled are different names for the same resource, discarded water that is recovered and *reused* for an intended purpose instead of being abandoned. Reclaimed and recycled water is an affordable, accessible, drought-proof water supply that supplements or replaces a community's nonpotable demand and can augment a water supply aquifer or storage reservoir for indirect beneficial uses. Eventually, reclaimed and recycled water may be used directly for public water system supplies. Reclaimed and recycled water receives high-level treatment, meets both traditional wastewater and drinking water quality requirements, and is required to provide a much higher level of reliability and redundancy. This water also is stored and distributed in the same manner as drinking water. Reclaimed and recycled water represents a major community investment, not only in capital expenses and higher operation and maintenance costs, but also in community involvement. Often, communities that invest in developing a reclaimed or recycled water utility are located in a drought-prone area where water is scarce.

2.0 UNIQUE PARAMETERS

As a hybrid of drinking water and wastewater, reclaimed and recycled water uses design, operation, and administration practices from both industries. There are, however, treatment processes, beneficial uses, and monitoring approaches that are used in unique ways. The following parameters are not specific to reclaimed and recycled water but are used in unique ways.

TABLE 7.1   Chemical and microbial concentrations.

Parameter	SI	Comments
Microconstituent concentration	ng/L or pg/L	Nanogram per liter or picogram per liter, drinking water (soluble organic carbon/volatile organic carbon) or wastewater (priority pollutant); commonly reported mg/L or µg/L
Bacterial concentration including total coliform, fecal coliform, *E. coli*, and other specific bacterial species	MPN/100 mL or CFU/100 mL	Dependent on test method chosen by laboratory
Virus concentration	PFU	PFU = plaque-forming units

TABLE 7.2   Pipeline separation.

Parameter	SI	Comments
Horizontal separation, S_h	m	Horizontal separation between parallel buried pipe
Vertical separation, S_v	m	Vertical separation, pipe crown to invert
Pipe diameter, D	m	Diameter of parallel buried pipe
Side cover, X	m	Distance from face of excavated trench to face of buried pipe
Critical trench depth, Z	m	Trench wall excavation depth at which native soil will stand in a vertical cut without sloughing
Cohesive strength, C	kg/m^2	Force per unit area from soil mechanics
Unit soil weight, γ	kg/m^2	Pounds per cubic foot
Soil friction angle, φ	m/m	Radians, angle of internal friction or angle of repose of soils

TABLE 7.3   Irrigation.

Parameter	SI	Comments
Evapotranspiration rate, ET	mm/day	Daily water crop water requirement
Net irrigation water requirement (NIW)	mm, L/m^2	Water volume applied per unit area
Total irrigation requirement	mm	NIW/field irrigation efficiency
Effective rainfall, Re	mm	Effective precipitation per month
Leaching requirement, LR	mm	
Salinity of applied water, ECw	mmho/cm	
Soil salinity tolerated by crop, ECc	mmho/cm	
Miscellaneous water requirements, Wm	mm	Volume required for frost protection, etc.
Water holding capacity (soils), Whc	mm/m	depth of water applied per unit depth of soil
Irrigation application rate, IR	mm/d	
Sodium adsorption ratio, SAR		$SAR = \dfrac{Na}{\sqrt{Ca + Mg/2}}$ Elements expressed in meq/L
Irrigable area, A_{irr}	ha	
Field irrigation *efficiency, e*	%	Ratio of plant water requirements to volume applied
Carrying capacity of soil, CCap	d	50% of plant-available soil water
Irrigation cycle, Nd	d	Number of days between irrigation applications

TABLE 7.4   Groundwater recharge and recovery.

Parameter	SI	Comments
Hydraulic head, H	M	
Pressure head, ψ	M	
Elevation head, Z	M	
Fluid pressure, P	N/m^2, Pa	
Fluid potential, Φ	m^2/s^2	
Mass density, ρ	kg/m^3	
Weight density, γ	unitless	Freeze and Cherry (1979)
Specific discharge, V	m/s	
Hydraulic conductivity, K	m/s	

TABLE 7.5   Water rights and access.

Parameter	SI	Comments
Annual withdrawal volume, Qa	m^3/d	Limits imposed by regulatory body
Instantaneous withdrawal, Qc	m^3/d, L/s	Pump or weir/gate capacity

TABLE 7.6   Reclaimed or recycled water demands and storage requirements.

Parameter	SI	Comments
Average day demand	L/d·ERU	Average daily water demand per user; ERU = equivalent residential user
Maximum day demand	L/d·ERU	Maximum daily water demand per user
Peak hour demand	L/m	Peak hour demand on distribution system
Fire flow requirement, FF	L/m	Minimum flow required for fire fighting
Flow duration, T_{ff}	h	Minimum time required to maintain minimum fire flow
Operating storage	L	Storage to account from periods of small demand such as pump operation
Equalizing storage, ES	L	Storage to account for peak demands
Standby storage, SS	L	Average daily storage requirement
Fire storage, FS	L	Fire storage = FF × T_{ff}
Dead storage, DS	L	Ineffective storage
Total storage, TS	L	TS = OS + ES + SS + FS + DS

References and Suggested Readings

American National Metric Council (1990) *Metric Editorial Guide*, 4th ed. (revised); American National Metric Council: Washington, D.C.

Beals, M.; Gross, L.; Harrell S. (1999) Permeability of Molecules, http://www.tiem.utk.edu/bioed/webmodules/permeability.htm (accessed May 2010).

Bolton, W. (2006) *Programmable Logic Controllers*, 4th ed.; Elsevier: Burlington, Massachusetts.

Camara, J. A. (2007) *Electrical Engineering Reference Manual for the Electrical and Computer PE Exam*, 7th ed.; Professional Publications Incorporated: Belmont, California.

Davis, M. (2011) *Water and Wastewater Engineering Design Principles and Practice*; McGraw-Hill: New York.

Davis, M. L.; Cornwell, D. A. (2008) *Introduction to Environmental Engineering*, 4th ed.; McGraw-Hill: New York.

Freeze, R. A.; Cherry, J. A. (1979) *Groundwater*; Prentice-Hall, Inc.: Upper Saddle River, New Jersey.

Hammer, M. J.; Hammer, M. J., Jr. (2008) *Water and Wastewater Technology*, 6th ed.; Pearson Prentice Hall, Inc.: Upper Saddle River, New Jersey.

International Bureau of Weights and Measures (*Bureau International des Poids et Mesures*) (2006) *The International System of Units (SI)*, 8th ed.; BIPM Pavillon de Breteuil: Sèvres Cedex, France.

Liptak, B. G. (Ed.) (2006) *Instrument Engineers' Handbook Fourth Edition, Process Control and Optimization*; Instrumentation Systems and Automation Society; CRC Press: Boca Raton, Florida.

McGhee, T. J. (1991) *Water Supply and Sewerage*, 6th Ed.; McGraw Hill: New York.

Metcalf and Eddy, Inc. (2003) *Wastewater Engineering: Treatment and Reuse*, 4th ed.; McGraw-Hill: New York.

Petruzella, F. D. (2005) *Programmable Logic Controllers*, 3rd ed.; McGraw-Hill, New York.

Sawyer, C. N.; McCarty, P. L.; Parkin, G. F. (1994) *Chemistry for Environmental Engineering*, 4th ed.; McGraw-Hill: New York.

Stumm, W.; Morgan, J. J. (1996) *Aquatic Chemistry: Chemical Equilibria and Rates in Natural Waters*, 3rd ed.; Wiley & Sons: New York.

Thompson, A., Jr.; Taylor, B. N. (2008) *Guide for the Use of the International System of Units (SI)*, Special Publication 811; National Institute of Standards and Technology: Gaithersburg, Maryland.

Water Environment Federation (2004) *Control of Odors and Emissions from Wastewater Treatment Plants*, Manual of Practice No. 25; McGraw-Hill: New York.

Water Environment Federation (2006) *Automation of Wastewater Treatment Facilities*, 3rd ed.; Manual of Practice No. 21; McGraw-Hill: New York.

Water Environment Federation (2009) *Wastewater Treatment Operator Training Manual, Wastewater Treatment: Overview, Preliminary Treatment, Primary Treatment, Natural Treatment Systems*; Water Environment Federation: Alexandria, Virginia.

Water Environment Federation; American Society of Civil Engineers; Environmental and Water Resources Institute (2009) *Design of Municipal Wastewater Treatment Plants*, 5th ed.; Manual of Practice No. 8; ASCE Manual of Practice and Report on Engineering No. 76; McGraw-Hill: New York.

Appendix A

Conversion Table for Acceptable Units

TABLE A.1 Conversion factors for acceptable units.

SI	Factor[a]	Customary unit
Â	1	C/s
atm	101 325	Pa (Pascal)
	1.013 25	bar
	760	Torr
	14.696	psi
C	1	As
	6.242E-18	electrons
cm	3.281E-02	ft
	0.393 70	in.
cm/h	0.3937	in./hr
g	0.0353	oz
	0.002 20	lb
$g/m^3 \cdot s$	0.2247	lb/hr/cu ft
$g/(m^2 \cdot s)$	17.70	lb/d/sq ft
	0.7373	lb/hr/sq ft
$g/(m^3 \cdot d)$	2.720	lb/d/ac-ft
g/kg	1.000E-03	lb/lb
	2.000	lb/ton
g/L	0.1335	oz/gal
g/m^3	5.841E-02	grains per gallon
	8.344	lb/mil. gal
	6.243E-02	lb/1000 cu ft
	8.347	lb/1000 gal
g/s	7.937	lb/hr
ha	2.471	ac

(continued)

TABLE A.1   Conversion factors for acceptable units (*continued*).

SI	Factor[a]	Customary unit
Hz	1.000	cycles/second
J/m^3	1.051E-03	kWh/mil. gal
Joules	0.2388	calorie (international)
	0.2389	calorie 15 °C
	0.2391	calorie 20 °C
	1.000E+07	Erg
	6.242E+20	eV
	23.73	ft-lb
	2.777E-04	W-hr
kg	2.205	lb (mass)
	1.102E-03	ton
	6.854E-02	slug
$kg/(m^2 \cdot h)$	0.2048	lb/sq ft/hr
kg/d	2.205	lb/d
kg/h	2.205	lb/hr
kg/ha	0.1	g/m^2
kg/ha·d	0.8922	lb/ac/d
kg/ha·yr	0.8922	lb/ac/yr
kg/L	3.613E-02	lb/cu in.
	8.347	lb/gal
	8.347E+06	lb/mil. gal
kg/m^2	0.2048	psf (mass dose)
$kg/m^2 \cdot d$	0.2048	lb/sq ft/hr
$kg/m^2 \cdot yr$	0.2048	lb/sq ft/yr
$kg/m^2 \cdot h$	0.2048	lb/sq ft/hr
$kg/m^2 \cdot s$	1.770E+04	lb/sq ft/d
	0.2048	lb/sq ft/s
	7.710E+08	lb/ac/d
kg/m^3	6.243E-02	lb/cu ft
	1.686	lb/cu yd
	8347	lb/mil. gal
	62.42	lb/1000 cu ft
$kg/m^3 \cdot d$	2725	lb/ac ft
	6.242E-02	lb/cu ft
	62.42	lb/1000 cu ft
kJ	0.9478	Btu (British thermal unit)
kJ/kg	0.429 92	Btu/lb
	859.8	Btu/ton
	0.2520	kWh/ton
	0.1260	kWh/1000 lb
	1.689E-04	hp-hr/lb

(*continued*)

TABLE **A.1** Conversion factors for acceptable units (*continued*).

SI	Factor[a]	Customary unit
kJ/m^2	1000	kg/s^2
kJ/m^3	2.68E-02	Btu/cu ft
	1.051	kWh/mil. gal
	1.408E-03	hp-hr/1000 gal
km	0.6214	mile
	0.5399	nautical mile
kN	0.2248	kip
kN/m^2	1000	Pa
$kN/m^3/kN/m^3$		
kPa	9.872E-03	atm
	20.89	psf (force)
	0.1450	psi (force)
	4.019	in. H_2O
	0.2961	in. Hg
	2.089E-02	kip/sq ft
kW	1.341	hp
L	3.532E-02	cu ft
	0.2642	gal
	2.114	pint
	1.057	quart
L/cap·d	0.2642	gpd/cap
L/d	4.087 345 69E-07	cfs
L/h	9.809 629 64E-06	cfs
L/m	452.2	gal/mi
L/m^2	0.0204	gal/sq ft
L/m^2·s	11.82	cfh/sq ft
	2.121E+03	gpd/sq ft
	9.237E+03	gpd/ac
	1.473	gpm/sq ft
L/min	0.0353	cfm
L/s	127.1	cfh
	2.119	cfm
	3.532E-02	cfs
	2.282E+04	gpd
	15.85	gpm
	2.283E-02	mgd
Lm^2·s	11.82	cfh/sq ft
	2.121E+03	gpd/sq ft
	9.237E+03	gpd/ac
	1.473	gpm/sq ft

(continued)

TABLE A.1　Conversion factors for acceptable units (*continued*).

SI	Factor[a]	Customary unit
m	1.000E+10	angstrom (A)
	0.5468	ftm (fathom)
	3.281	ft
	3.281	ft H_2O
	39.37	in.
	1.094	yd
m/h	3.281	ft/hr
m/s	3.281	ft/sec
	1.181E+04	ft/hr
	196.9	fpm
	2.238	mph
m/s^2	3.281	ft/sec^2
m^2	2.471E-04	ac
	10.76	sq ft
	1.550E+03	sq in.
	1.196	sq yd
m^2/g		
m^2/m^3	3.048	sq ft/cu ft
m^3	35.32	cu ft
	8.107 13E-04	ac-ft
$m^3/(m^2 \cdot d)$	24.55	gpd/sq ft
	1.704E-02	gpm/sq ft
$m^3/(m^2 \cdot h)$	3.281	cfh/sq ft
$m^3/(m^2 \cdot s)$	1.429E+05	cfs/ac
	9.149E+07	cfs/sq mi
	9.238E+10	gpd/ac
	2.121E+06	gpd/sq ft
	1.473E+03	gpm/sq ft
	9.238E+04	mgd/ac
	1.181E+04	cfh/sq ft
m^3/d	0.1835	gpm
	2.642E-04	mgd
	2.642E+02	gpd
m^3/h		
$m^3/ha \cdot d$	106.9	gpd/ac
	1.069E-04	mgd/ac
$m^3/m^2 \cdot d$	24.55	gpd/sq ft
	1.704E-02	gpm/sq ft
m^3/m^3	0.1337	cu ft/gal
	1.337E+05	cu ft/mil. gal
		(*continued*)

TABLE A.1   Conversion factors for acceptable units (*continued*).

SI	Factor[a]	Customary unit
m^3/s	1.271E+05	cfh or cu ft/hr
	2.119E+03	cfm
	35.32	cfs
	9.515E+05	gph
	1.585E+04	gpm
	22.83	mgd
Mg	1.102	ton (short)
	0.9843	ton (long)
Mg/ha	0.4460	ton/ac
mg/L	5.841E-02	grains per gallon
	1.00	ppm
ML	0.2642	mil. gal
ML/d	0.2642	mgd
mm	3.937E-02	in.
mm/h	3.937E-02	in./hr
mm/s	0.1969	fpm
	11.81	ft/hr
MPN/L	0.1000	MPN/100 mL
nm	1.00E-09	m
Pa	0.00750	Torr
	9.8692E-06	atm
W	3.412	Btu/hr
	947.8	Btu/sec
	0.2388	cal/sec
	44.27	ft-lb/min
	0.7380	ft-lb/sec
	1.341E-03	hp
	2.400E-02	kWh/d

[a]Multiply SI unit by conversion factor to obtain the customary unit. Divide the customary unit by the conversion factor to obtain SI unit.

Appendix B

Acronyms

ADD	Average day demand
BOD	Biochemical oxygen demand
COD	Chemical oxygen demand
CFU	Colony forming units
CUR	Carbon use rate
DAF	Dissolved air flotation
D/T	Dilution to threshold
ERU	Equivalent residential user
FCU	Fecal coliform units
F/M	Food-to-microorganism ratio
GAC	Granular activated carbon
GBT	Gravity belt thickener
HRT	Hydraulic retention time
HLR	Hydraulic loading rate
IFAS	Integrated fixed-film activated sludge
LOX	Liquid oxygen
MBBR	Moving-bed biofilm reactor
MBR	Membrane bioreactor
MDD	Maximum day demand
MLSS	Mixed liquor suspended solids
N	Newtons, as a pressure unit
N	Normal, as a unit prefix
N	Normality, when referring to chemicals
NDMA	N-nitrosodimethylamine
NPSH	Net pump suction head
NTU	Nephelometric turbidity units

ORP	Oxidation–reduction potential
PFU	Plaque-forming units
PHD	Peak-hour demand
RDT	Rotary-drum thickener
SCCR	Short-circuit current rating
SLR	Solids loading rate
SOR	Surface overflow rate
SOTE	Standard oxygen-transfer efficiency
SRT	Solids retention time
SVI	Sludge volume index
TIR	Total irrigation requirement
TKN	Total Kjeldahl nitrogen
TMP	Transmembrane pressure
TSS	Total suspended solids
VFA	Volatile fatty acids
VSS	Volatile suspended solids